Operating System Security

Synthesis Lectures on Information Security, Privacy and Trust

Editor
Ravi Sandhu, University of Texas, San Antonio

Operating System Security
Trent Jaeger
2008

Operating System Security

Trent Jaeger

www.morganclaypool.com

ISBN: 9781598292121   paperback
ISBN: 9781598292138   ebook

DOI 10.2200/S00126ED1V01Y200808SPT001

A Publication in the Morgan & Claypool Publishers series
SYNTHESIS LECTURES ON INFORMATION SECURITY, PRIVACY AND TRUST

Lecture #1
Series Editor: Ravi Sandhu, University of Texas, San Antonio

Series ISSN
Synthesis Lectures on Information Security, Privacy and Trust
ISSN pending.

Operating System Security

Trent Jaeger
The Pennsylvania State University

SYNTHESIS LECTURES ON INFORMATION SECURITY, PRIVACY AND TRUST #1

MORGAN &CLAYPOOL PUBLISHERS

ABSTRACT

Operating systems provide the fundamental mechanisms for securing computer processing. Since the 1960s, operating systems designers have explored how to build "secure" operating systems — operating systems whose mechanisms protect the system against a motivated adversary. Recently, the importance of ensuring such security has become a mainstream issue for all operating systems. In this book, we examine past research that outlines the requirements for a secure operating system and research that implements example systems that aim for such requirements. For system designs that aimed to satisfy these requirements, we see that the complexity of software systems often results in implementation challenges that we are still exploring to this day. However, if a system design does not aim for achieving the secure operating system requirements, then its security features fail to protect the system in a myriad of ways. We also study systems that have been retrofit with secure operating system features after an initial deployment. In all cases, the conflict between function on one hand and security on the other leads to difficult choices and the potential for unwise compromises. From this book, we hope that systems designers and implementors will learn the requirements for operating systems that effectively enforce security and will better understand how to manage the balance between function and security.

KEYWORDS

Operating systems, reference monitor, mandatory access control, secrecy, integrity, virtual machines, security kernels, capabilities, access control lists, multilevel security, policy lattice, assurance

To Dana, Alec, and David for their love and support

Contents

Preface

Operating system security forms the foundation of the secure operation of computer systems. In this book, we define what is required for an operating system to ensure enforcement of system security goals and evaluate how several operating systems have approached such requirements.

WHAT THIS BOOK IS ABOUT

Chapter	Topic
2. Fundamentals	Define an Ideal, Secure OS
3. Multics	The First OS Designed for Security Goals
4. Ordinary OS's	Why Commercial OS's Are Not Secure
5. Verifiable Security	Define Precise Security Goals
6. Security Kernels	Minimize OS's Trusted Computing Base
7. Secure Commercial OS's	Retrofit Security into Commercial OS's
8. Solaris Trusted Extensions Case Study	MLS Extension of Solaris OS
9. SELinux Case Study	Examine Retrofit of Linux Specifically
10. Capability Systems	Ensure Security Goal Enforcement
11. Virtual Machines	Identify Necessary Security Mechanisms
12. System Assurance	Methodologies to Verify Correct Enforcement

Figure 1: Overview of the Chapters in this book.

In this book, we examine what it takes to build a secure operating system, and explore the major systems development approaches that have been applied towards building secure operating systems. This journey has several goals shown in Figure 1. First, we describe the fundamental concepts and mechanisms for enforcing security and define secure operating systems (Chapter 2). Second, we examine early work in operating systems to show that it may be possible to build systems that approach a secure operating system, but that ordinary, commercial operating systems are not secure fundamentally (Chapters 3 and 4, respectively). We next describe the formal security goals and corresponding security models proposed for secure operating systems (Chapter 5). We then survey a variety of approaches applied to the development of secure operating systems (Chapters 6 to 11). Finally, we conclude with a discussion of system assurance methodologies (Chapter 12).

The first half of the book (Chapters 2 to 5) aims to motivate the challenges of building a secure operating system. Operating systems security is so complex and broad a subject that we cannot introduce everything without considering some examples up front. Thus, we start with just the fundamental concepts and mechanisms necessary to understand the examples. Also, we take the step of showing what a system designed to the secure operating system definition (i.e., Multics in Chapter 3) looks like and what insecure operating systems (i.e., UNIX and Windows in Chapter 4) looks like and why. In Chapter 5, we then describe concrete security goals and how they can be expressed once the reader has an understanding of what is necessary to secure a system.

The second half of the book surveys the major, distinct approaches to building secure operating systems in Chapters 6 to 11. Each of the chapters focuses on the features that are most important to these approaches. As a result, each of these chapters has a different emphasis. For example, Chapter 6 describes security kernel systems where the operating system is minimized and leverages hardware features and low-level system mechanisms. Thus, this chapter describes the impact of hardware features and the management of hardware access on our ability to construct effective and flexible secure operating systems. Chapter 7 summarizes a variety of ways that commercial operating systems have been extended with security features. Chapters 8 and 9 focus on retrofitting security features on existing, commercial operating systems, Solaris and Linux, respectively. Glenn Faden and Christoph Schuba from Sun Microsystems detail the Solaris (TM) Trusted Extensions. In these chapters, the challenges include modifying the system architecture and policy model to enforce security goals. Here, we examine adding security to user-level services, and extending security enforcement into the network. The other chapters examine secure capability systems and how capability semantics are made secure (Chapter 10) and secure virtual machine systems to examine the impact and challenges of using virtualization to improve security (Chapter 11).

The book concludes with the chapter on system assurance (Chapter 12). In this chapter, we discuss the methodologies that have been proposed to verify that a system is truly secure. Assurance verification is a major requirement of secure operating systems, but it is still at best a semi-formal process, and in practice an informal process for general-purpose systems.

The contents of this book derive from the work of many people over many years. Building an operating system is a major project, so it is not surprising that large corporate and/or research teams are responsible for most of the operating systems in this book. However, several individual researchers have devoted their careers to operating systems security, so they reappear throughout the book in various projects advancing our knowledge on the subject. We hope that their efforts inspire future researchers to tackle the challenges of improving operating systems security.

WHAT THIS BOOK IS NOT ABOUT

As with any book, the scope of investigation is limited and there are many related and supporting efforts that are not described. Some operating system development approaches and several representative operating systems are not detailed in the book. While we attempted to include all broad approaches to building secure systems, some may not quite fit the categorizations and there are several systems that have interesting features that could not be covered in depth.

Other operating systems problems appear to be related to security, but are outside the scope of this book. For example, *fault tolerance* is the study of how to maintain the correctness of a computation given the failure of one or more components. Security mechanisms focus on ensuring that security goals are achieved regardless of the behavior of a process, so fault tolerance would depend on security mechanisms to be able to resurrect or maintain a computation. The area of *survivability* is also related, but it involves fault tolerance in the face of catastrophic failures or natural disasters. Its goals also depend on effective computer security.

There are also several areas of computer science whose advances may benefit operating system security, but which we omit in this book. For example, recent advances in *source code analysis* improves the correctness of system implementations by identifying bugs [82, 209, 49] and even being capable of proving certain properties of small programs, such as device drivers [210, 18]. Further, programming languages that enable verifiable enforcement of security properties, such as *security-typed languages* [219, 291], also would seem to be necessary to ensure that all the trusted computing base's code enforces the necessary security goals. In general, we believe that improvements in languages, programming tools for security, and analysis of programs for security are necessary to verify the requirements of secure operating systems.

Also, a variety of programs also provide security mechanisms. Most notably, these include databases (e.g., Oracle) and application-level virtual machines (e.g., Java). Such programs are only relevant to the construction of a secure operating system if they are part of the trusted computing base. As this is typically not the case, we do not discuss these application-level mechanisms.

Ultimately, we hope that the reader gains a clearer understanding of the challenging problem of building a secure operating system and an appreciation for the variety of solutions applied over the years. Many past and current efforts have explored these challenges in a variety of ways. We hope that the knowledge and experiences of the many people whose work is captured in this book will serve as a basis for comprehensive and coherent security enforcement in the near future.

Trent Jaeger
The Pennsylvania State University
August 2008

CHAPTER 1

Introduction

Operating systems are the software that provides access to the various hardware resources (e.g., CPU, memory, and devices) that comprise a computer system as shown in Figure 1.1. Any program that is run on a computer system has instructions executed by that computer's CPU, but these programs may also require the use of other peripheral resources of these complex systems. Consider a program that allows a user to enter her password. The operating system provides access to the disk device on which the program is stored, access to device memory to load the program so that it may be executed, the display device to show the user how to enter her password, and keyboard and mouse devices for the user to enter her password. Of course, there are now a multitude of such devices that can be used seamlessly, for the most part, thanks to the function of operating systems.

As shown in Figure 1.1, operating systems run programs in *processes*. The challenge for an operating system developer is to permit multiple concurrently executing processes to use these resources in a manner that preserves the independence of these processes while providing fair sharing of these resources. Originally, operating systems only permitted one process to be run at a time (e.g., *batch systems*), but as early as 1960, it became apparent that computer productivity would be greatly enhanced by being able to run multiple processes concurrently [87]. By *concurrently*, we mean that while only one process uses a computer's CPU at a time, multiple other processes may be in various states of execution at the same time, and the operating system must ensure that these executions are performed effectively. For example, while the computer waits for a user to enter her password, other processes may be run and access system devices as well, such as the network. These systems were originally called *timesharing systems*, but they are our default operating systems today.

To build any successful operating system, we identify three major tasks. First, the operating system must provide various mechanisms that enable high performance use of computer resources. Operating systems must provide efficient *resource mechanisms*, such as file systems, memory management systems, network protocol stacks, etc., that define how processes use the hardware resources. Second, it is the operating system's responsibility to switch among the processes fairly, such that the user experiences good performance from each process in concert with access to the computer's devices. This second problem is one of *scheduling* access to computer resources. Third, access to resources should be controlled, such that one process cannot inadvertently or maliciously impact the execution of another. This third problem is the problem of ensuring the *security* of all processes run on the system.

Ensuring the secure execution of all processes depends on the correct implementation of resource and scheduling mechanisms. First, any correct resource mechanism must provide boundaries between its objects and ensure that its operations do not interfere with one another. For example, a file system must not allow a process request to access one file to overwrite the disk space allocated

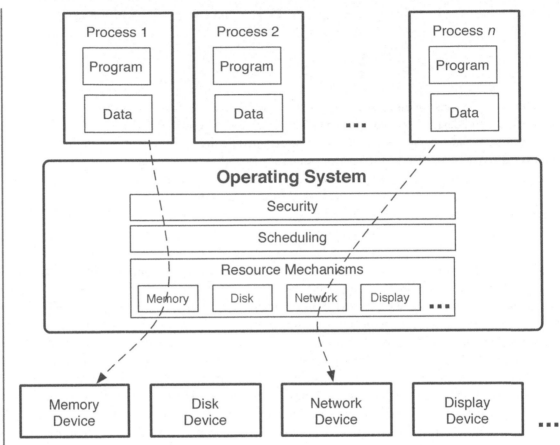

Figure 1.1: An operating system runs *security*, *scheduling*, and *resource mechanisms* to provide *processes* with access to the computer system's resources (e.g., CPU, memory, and devices).

to another file. Also, file systems must ensure that one write operation is not impacted by the data being read or written in another operation. Second, scheduling mechanisms must ensure availability of resources to processes to prevent denial of service attacks. For example, the algorithms applied by scheduling mechanisms must ensure that all processes are eventually scheduled for execution. These requirements are fundamental to operating system mechanisms, and are assumed to be provided in the context of this book. The scope of this book covers the misuse of these mechanisms to inadvertently or, especially, maliciously impact the execution of another process.

Security becomes an issue because processes in modern computer systems interact in a variety of ways, and the sharing of data among users is a fundamental use of computer systems. First, the output of one process may be used by other processes. For example, a programmer uses an editor program to write a computer program's source code, compilers and linkers to transform the program code into a form in which it can be executed, and debuggers to view the executing processes image to find errors in source code. In addition, a major use of computer systems is to share information with other users. With the ubiquity of Internet-scale sharing mechanisms, such as e-mail, the web, and instant messaging, users may share anything with anyone in the world. Unfortunately, lots of people, or at least lots of email addresses, web sites, and network requests, want to share stuff with you that aims to circumvent operating system security mechanisms and cause your computer to share additional, unexpected resources. The ease with which malware can be conveyed and the variety of ways that users and their processes may be tricked into running malware present modern operating system developers with significant challenges in ensuring the security of their system's execution.

The challenge in developing operating systems security is to design security mechanisms that protect process execution and their generated data in an environment with such complex interactions. As we will see, formal security mechanisms that enforce provable security goals have been defined, but these mechanisms do not account or only partially account for the complexity of practical systems. As such, the current state of operating systems security takes two forms: (1) constrained systems that can enforce security goals with a high degree of assurance and (2) general-purpose systems that can enforce limited security goals with a low to medium degree of assurance. First, several systems have been developed over the years that have been carefully crafted to ensure correct (i.e., within some low tolerance for bugs) enforcement of specific security goals. These systems generally support few applications, and these applications often have limited functionality and lower performance requirements. That is, in these systems, security is the top priority, and this focus enables the system developers to write software that approaches the ideal of the formal security mechanisms mentioned above. Second, the computing community at large has focused on function and flexibility, resulting in general-purpose, extensible systems that are very difficult to secure. Such systems are crafted to simplify development and deployment while achieving high performance, and their applications are built to be feature-rich and easy to use. Such systems present several challenges to security practitioners, such as insecure interfaces, dependence of security on arbitrary software, complex interaction with untrusted parties anywhere in the world, etc. But, these systems have defined how the user community works with computers. As a result, the security community faces a difficult task for ensuring security goals in such an environment.

However, recent advances are improving both the utility of the constrained systems and the security of the general-purpose systems. We are encouraged by this movement, which is motivated by the general need for security in all systems, and this book aims to capture many of the efforts in building security into operating systems, both constrained and general-purpose systems, with the aim of enabling broader deployment and use of security function in future operating systems.

1.1   SECURE OPERATING SYSTEMS

The ideal goal of operating system security is the development of a secure operating system. *A secure operating system provides security mechanisms that ensure that the system's security goals are enforced despite the threats faced by the system.* These security mechanisms are designed to provide such a guarantee in the context of the resource and scheduling mechanisms. Security goals define the requirements of secure operation for a system for any processes that it may execute. The security mechanisms must ensure these goals regardless of the possible ways that the system may be misused (i.e., is threatened) by attackers.

The term "secure operating system" is both considered an ideal and an oxymoron. Systems that provide a high degree of assurance in enforcement have been called secure systems, or even more frequently "trusted" systems [1]. However, it is also true that no system of modern complexity is completely secure. The difficulty of preventing errors in programming and the challenges of trying to remove such errors means that no system as complex as an operating system can be completely secure.

Nonetheless, we believe that studying how to build an ideal secure operating system to be useful in assessing operating systems security. In Chapter 2, we develop a definition of *secure operating system* that we will use to assess several operating systems security approaches and specific implementations of those approaches. While no implementation completely satisfies this ideal definition, its use identifies the challenges in implementing operating systems that satisfy this ideal in practice. The aim is multi-fold. First, we want to understand the basic strengths of common security approaches. Second, we want to discover the challenges inherent to each of these approaches. These challenges often result in difficult choices in practical application. Third, we want to study the application of these approaches in practical environments to evaluate the effectiveness of these approaches to satisfy the ideal in practice. While it appears impractical to build an operating system that satisfies the ideal definition, we hope that studying these systems and their security approaches against the ideal will provide insights that enable the development of more effective security mechanisms in the future.

To return to the general definition of a secure operating system from the beginning of this section, we examine the general requirements of a secure operating system. To build any secure system requires that we consider how the system achieves its *security goals* under a set of threats (i.e., a *threat model*) and given a set of software, including the security mechanisms, that must be trusted [2] (i.e., a *trust model*).

1.2   SECURITY GOALS

A security goal defines the operations that can be executed by a system while still preventing unauthorized access. It should be defined at a high-level of abstraction, not unlike the way that an algorithm's worst-case complexity prescribes the set of implementations that satisfy that requirement. A security goal defines a requirement that the system's design can satisfy (e.g., the way pseudocode can be proven to fulfill the complexity requirement) and that a correct implementation must fulfill (e.g., the way that an implementation can be proven experimentally to observe the complexity).

---

[1] For example, the first description of criteria to verify that a system implements correct security mechanisms is called the Trusted Computer System Evaluation Criteria [304].

[2] We assume that hardware is trusted to behave as expected. Although the hardware devices may have bugs, the trust model that we will use throughout this book assumes that no such bugs are present.

Security goals describe how the system implements accesses to system resources that satisfy the following: *secrecy*, *integrity*, and *availability*. A system access is traditionally stated in terms of which *subjects* (e.g., processes and users) can perform which *operations* (e.g., read and write) on which *objects* (e.g., files and sockets). Secrecy requirements limit the objects that individual subjects can *read* because objects may contain secrets that not all subjects are permitted to know. Integrity requirements limit the objects that subjects can *write* because objects may contain information that other subjects *depend on* for their correct operation. Some subjects may not be trusted to modify those objects. Availability requirements limit the system resources (e.g., storage and CPU) that subjects may *consume* because they may exhaust these resources. Much of the focus in secure operating systems is on secrecy and integrity requirements, although availability may indirectly impact these goals as well.

The security community has identified a variety of different security goals. Some security goals are defined in terms of security requirements (i.e., secrecy and integrity), but others are defined in terms of function, in particular ways to limit function to improve security. An example of a goal defined in terms of security requirements is the *simple-security property* of the Bell-LaPadula model [23]. This goal states that a process cannot read an object whose secrecy classification is higher than the process's. This goal limits operations based on a security requirement, secrecy. An example of an functional security goal is the *principle of least privilege* [265], which limits a process to only the set of operations necessary for its execution. This goal is functional because it does not ensure that the secrecy and/or integrity of a system is enforced, but it encourages functional restrictions that may prevent some attacks. However, we cannot prove the absence of a vulnerability using functional security goals. We discuss this topic in detail in Chapter 5.

The task of the secure operating system developer is to define security goals for which the security of the system can be verified, so functional goals are insufficient. On the other hand, secrecy and integrity goals prevent function in favor of security, so they may be too restrictive for some production software. In the past, operating systems that enforced secrecy and integrity goals (i.e., the constrained systems above) were not widely used because they precluded the execution of too many applications (or simply lacked popular applications). Emerging technology, such as virtual machine technology (see Chapter 11), enables multiple, commercial software systems to be run in an isolated manner on the same hardware. Thus, software that used to be run on the same system can be run in separate, isolated virtual systems. It remains to be seen whether such isolation can be leveraged to improve system security effectively. Also, several general-purpose operating systems are now capable of expressing and enforcing security goals. Whether these general-purpose systems will be capable of implementing security goals or providing sufficient assurance for enforcing such goals is unclear. However, in either case, security goals must be defined and a practical approach for enforcing such goals, that enables the execution of most popular software in reasonable ways, must be identified.

1.3   TRUST MODEL

A system's *trust model* defines the set of software and data upon which the system depends for correct enforcement of system security goals. For an operating system, its trust model is synonymous with the system's *trusted computing base* (TCB).

Ideally, a system TCB should consist of the minimal amount of software necessary to enforce the security goals correctly. The software that must be trusted includes the software that defines the security goals and the software that enforces the security goals (i.e., the operating system's security mechanism). Further, software that bootstraps this software must also be trusted. Thus, an ideal TCB would consist of a bootstrapping mechanism that enables the security goals to be loaded and subsequently enforced for lifetime of the system.

In practice, a system TCB consists of a wide variety of software. Fundamentally, the enforcement mechanism is run within the operating system. As there are no protection boundaries between operating system functions (i.e., in the typical case of a monolithic operating system), the enforcement mechanism must trust all the operating system code, so it is part of the TCB.

Further, a variety of other software running outside the operating system must also be trusted. For example, the operating system depends on a variety of programs to authenticate the identity of users (e.g., `login` and `SSH`). Such programs must be trusted because correct enforcement of security goals depends on correct identification of users. Also, there are several services that the system must trust to ensure correct enforcement of security goals. For example, windowing systems, such as the X Window System [345], perform operations on behalf of all processes running on the operating system, and these systems provide mechanisms for sharing that may violate the system's security goals (e.g., cut-and-paste from one application to another) [85]. As a result, the X Window Systems and a variety of other software must be added to the system's TCB.

The secure operating system developer must prove that their systems have a viable trust model. This requires that: (1) the system TCB must mediate all security-sensitive operations; (2) verification of the correctness of the TCB software and its data; and (3) verification that the software's execution cannot be tampered by processes outside the TCB. First, identifying the TCB software itself is a nontrivial task for reasons discussed above. Second, verifying the correctness of TCB software is a complex task. For general-purpose systems, the amount of TCB software outside the operating system is greater than the operating system software that is impractical to verify formally. The level of trust in TCB software can vary from software that is formally-verified (partially), fully-tested, and reviewed to that which the user community trusts to perform its appointed tasks. While the former is greatly preferred, the latter is often the case. Third, the system must protect the TCB software and its data from modification by processes outside the TCB. That is, the integrity of the TCB must be protected from the threats to the system, described below. Otherwise, this software can be tampered, and is no longer trustworthy.

1.4   THREAT MODEL

A *threat model* defines a set of operations that an *attacker* may use to *compromise* a system. In this threat model, we assume a powerful attacker who is capable of injecting operations from the network and may be in control of some of the running software on the system (i.e., outside the trusted computing base). Further, we presume that the attacker is actively working to violate the system security goals. If an attacker is able to find a vulnerability in the system that provides access to secret information (i.e., violate secrecy goals) or permits the modification of information that subjects depend on (i.e., violate integrity goals), then the attacker is said to have compromised the system.

Since the attacker is actively working to violate the system security goals, we must assume that the attacker may try any and all operations that are permitted to the attacker. For example, if an attacker can only access the system via the network, then the attacker may try to send any operation to any processes that provide network access. Further, if an attacker is in control of a process running on the system, then the attacker will try any means available to that process to compromise system security goals.

This threat model exposes a fundamental weakness in commercial operating systems (e.g., UNIX and Windows); they assume that all software running on behalf of a subject is trusted by that subject. For example, a subject may run a word processor and an email client, and in commercial systems these processes are trusted to behave as the user would. However, in this threat model, both of these processes may actually be under the control of an attacker (e.g., via a document macro virus or via a malicious script or email attachment). Thus, a secure operating system cannot trust processes outside of the TCB to behave as expected. While this may seem obvious, commercial systems trust any user process to manage the access of that user's data (e.g., to change access rights to a user's files via chmod in a UNIX system). This can result in the leakage of that user's secrets and the modification of data that the user depends on.

The task of a secure operating system developer is to protect the TCB from the types of threats described above. Protecting the TCB ensures that the system security goals will always be enforced regardless of the behavior of user processes. Since user processes are untrusted, we cannot depend on them, but we can protect them from threats. For example, secure operating system can prevent a user process with access to secret data from leaking that data, by limiting the interactions of that process. However, protecting the TCB is more difficult because it interacts with a variety of untrusted processes. A secure operating system developer must identify such threats, assess their impact on system security, and provide effective countermeasures for such threats. For example, a trusted computing base component that processes network requests must identify where such untrusted requests are received from the network, determine how such threats can impact the component's behavior, and provide countermeasures, such as limiting the possible commands and inputs, to protect the component. The secure operating system developer must ensure that all the components of the trusted computing base prevent such threats correctly.

1.5   SUMMARY

While building a truly secure operating system may be infeasible, operating system security will improve immensely if security becomes a focus. To do so requires that operating systems be designed to enforce security goals, provide a clearly-identified trusted computing base that defines a trust model, define a threat model for the trusted computing base, and ensure protection of the trusted computing base under that model.

CHAPTER 2

Access Control Fundamentals

An *access enforcement mechanism* authorizes requests (e.g., system calls) from multiple *subjects* (e.g., users, processes, etc.) to perform *operations* (e.g., read, write, etc.) on objects (e.g., files, sockets, etc.). An operating system provides an access enforcement mechanism. In this chapter, we define the fundamental concepts of access control: a *protection system* that defines the access control specification and a *reference monitor* that is the system's access enforcement mechanism that enforces this specification. Based on these concepts, we provide an ideal definition for a secure operating system. We use that definition to evaluate the operating systems security of the various systems examined in this book.

2.1   PROTECTION SYSTEM

The security requirements of a operating system are defined in its *protection system*.

Definition 2.1.   A *protection system* consists of a *protection state*, which describes the operations that system subjects can perform on system objects, and a set of *protection state operations*, which enable modification of that state.

A protection system enables the definition and management of a protection state. A *protection state* consists of the specific system subjects, the specific system objects, and the operations that those subjects can perform on those objects. A protection system also defines *protection state operations* that enable a protection state to be modified. For example, protection state operations are necessary to add new system subjects or new system objects to the protection state.

2.1.1   LAMPSON'S ACCESS MATRIX
Lampson defined the idea that a protection state is represented by an *access matrix*, in general, [176].

Definition 2.2.   An *access matrix* consists of a set of subjects $s \in S$, a set of objects $o \in O$, a set of operations $op \in OP$, and a function $ops(s, o) \subseteq OP$, which determines the operations that subject s can perform on object o. The function $ops(s, o)$ is said to return a set of operations corresponding to cell (s, o).

Figure 2.1 shows an access matrix. The matrix is a two-dimensional representation where the set of subjects form one axis and the set of objects for the other axis. The cells of the access matrix store the operations that the corresponding subject can perform on the corresponding object. For example, subject Process 1 can perform read and write operations on object File 2.

	File 1	File 2	File 3	Process 1	Process 2
Process 1	Read	Read, Write	Read, Write	Read	-
Process 2	-	Read	Read, Write	-	Read

Figure 2.1: Lampson's Access Matrix

If the subjects correspond to processes and the objects correspond to files, then we need protection state operations to update the protection state as new files and processes are created. For example, when a new file is created, at least the creating process should gain access to the file. In this case, a protection state operation create_file(process, file) would add a new column for the new file and add read and write operations to the cell (*process*, *file*).

Lampson's access matrix model also defines operations that determine which subjects can modify cells. For example, Lampson defined an own operation that defines ownership operations for the associated object. When a subject is permitted for the own operation for an object *o*, that subject can modify the other cells associated with that object *o*. Lampson also explored delegation of ownership operations to other subjects, so others may manage the distribution of permissions.

The access matrix is used to define the *protection domain* of a process.

Definition 2.3.   A *protection domain* specifies the set of resources (objects) that a process can access and the operations that the process may use to access such resources.

By examining the rows in the access matrix, one can see all the operations that a subject is authorized to perform on system resources. This determines what information could be read and modified by a processes running on behalf of that subject. For a secure operating system, we will want to ensure that the protection domain of each process satisfies system security goals (e.g., secrecy and integrity).

A process at any time is associated with one or more subjects that define its protection domain. That is, the operations that it is authorized to perform are specified by one or more subjects. Systems that we use today, see Chapter 4, compose protection domains from a combination of subjects, including users, their groups, aliases, and ad hoc permissions. However, protection domains can also be constructed from an intersection of the associated subjects (e.g., Windows 2000 Restricted Contexts [303]). The reason to use an intersection of subjects' permissions is to restrict the protection domain to permissions shared by all, rather than giving the protection domain subjects extra permissions that they would not normally possess.

Because the access matrix would be a sparse data structure in practice (i.e., most of the cells would not have any operations), other representations of protection states are used in practice. One representation stores the protection state using individual object columns, describing which subjects have access to a particular object. This representation is called an *access control list* or ACL. The other representation stores the other dimension of the access matrix, the subject rows. In this case, the objects that a particular subject can access are stored. This representation is called a *capability list* or C-List.

There are advantages and disadvantages to both the C-List and ACL representations of protection states. For the ACL approach, the set of subjects and the operations that they can perform are stored with the objects, making it easy to tell which subjects can access an object at any time. Administration of permissions seems to be more intuitive, although we are not aware of any studies to this effect. C-Lists store the set of objects and operations that can be performed on them are stored with the subject, making it easy to identify a process's protection domain. The systems in use today, see Chapter 4, use ACL representations, but there are several systems that use C-Lists, as described in Chapter 10.

2.1.2   MANDATORY PROTECTION SYSTEMS

This access matrix model presents a problem for secure systems: untrusted processes can tamper with the protection system. Using protection state operations, untrusted user processes can modify the access matrix by adding new subjects, objects, or operations assigned to cells. Consider Figure 2.1. Suppose Process 1 has ownership over File 1. It can then grant any other process read or write (or potentially even ownership) access over File 1. A protection system that permits untrusted processes to modify the protection state is called a *discretionary access control* (DAC) system. This is because the protection state is at the discretion of the users and any untrusted processes that they may execute.

The problem of ensuring that particular protection state and all possible future protection states derivable from this state will not provide an unauthorized access is called the *safety problem* [130] [1]. It was found that this problem is undecidable for protection systems with compound protection state operations, such as for create_file above which both adds a file column and adds the operations to the owner's cell. As a result, it is not possible, in general, to verify that a protection state in such a system will be secure (i.e., satisfy security goals) in the future. To a secure operating system designer, such a protection system cannot be used because it is not tamperproof; an untrusted process can modify the protection state, and hence the security goals, enforced by the system.

We say that the protection system defined in Definition 2.1 aims to enforce the requirement of *protection*: one process is protected from the operations of another only if both processes behave benignly. If no user process is malicious, then with some degree of certainly, the protection state will still describe the true security goals of the system, even after several operations have modified the protection state. Suppose that a File 1 in Figure 2.1 stores a secret value, such as a private key in a public key pair [257], and File 2 stores a high integrity value like the corresponding public key. If Process 1 is non-malicious, then it is unlikely that it will leak the private key to Process 2 through either File 1 or File 2 or by changing the Process 2's permissions to File 1. However, if Process 1 is malicious, it is quite likely that the private key will be leaked. To ensure that the

---

[1]For a detailed analysis of the *safety problem* see Bishop's textbook [29].

secrecy of File 1 is enforced, all processes that have access to that file must not be able to leak the file through the permissions available to that process, including via protection state operations.

Similarly, the access matrix protection system does not ensure the integrity of the public key file File 2, either. In general, an attacker must not be able to modify any user's public key because this could enable the attacker to replace this public key with one whose private key is known to the attacker. Then, the attacker could masquerade as the user to others. Thus, the integrity compromise of File 2 also could have security ramifications. Clearly, the access matrix protection system cannot protect File 2 from a malicious Process 1, as it has write access to File 2. Further, a malicious Process 2 could enhance this attack by enabling the attacker to provide a particular value for the public key. Also, even if Process 1 is not malicious, a malicious Process 2 may be able to trick Process 1 into modifying File 2 in a malicious way depending on the interface and possible vulnerabilities in Process 1. Buffer overflow vulnerabilities are used in this manner for a malicious process (e.g., Process 2) to take over a vulnerable process (e.g., Process 1) and use its permissions in an unauthorized manner.

Unfortunately, the protection approach underlying the access matrix protection state is naive in today's world of malware and connectivity to ubiquitous network attackers. We see in Chapter 4 that today's computing systems are based on this protection approach, so they cannot be ensure enforcement of secrecy and integrity requirements. Protection systems that can enforce secrecy and integrity goals must enforce the requirement of *security: where a system's security mechanisms can enforce system security goals even when any of the software outside the trusted computing base may be malicious.* In such a system, the protection state must be defined based on the accurate identification of the secrecy and integrity of user data and processes, and no untrusted processes may be allowed to perform protection state operations. Thus, the dependence on potentially malicious software is removed, and a concrete basis for the enforcement of secrecy and integrity requirements is possible.

This motivates the definition of a mandatory protection system below.

Definition 2.4. A *mandatory protection system* is a protection system that can only be modified by trusted administrators via trusted software, consisting of the following state representations:

- A *mandatory protection state* is a protection state where subjects and objects are represented by *labels* where the state describes the operations that subject labels may take upon object labels;

- A *labeling state* for mapping processes and system resource objects to labels;

- A *transition state* that describes the legal ways that processes and system resource objects may be relabeled.

For secure operating systems, the subjects and objects in an access matrix are represented by system-defined *labels*. A label is simply an abstract identifier—the assignment of permissions to a label defines its security semantics. Labels are tamperproof because: (1) the set of labels is defined by trusted administrators using trusted software and (2) the set of labels is immutable. Trusted administrators define the access matrix's labels and set the operations that subjects of particular labels can perform on objects of particular labels. Such protection systems are *mandatory access control* (MAC) systems because the protection system is immutable to untrusted processes [2]. Since the set of labels cannot be changed by the execution of user processes, we can prove the security goals enforced by the access matrix and rely on these goals being enforced throughout the system's execution.

Of course, just because the set of labels are fixed does not mean that the set of processes and files are fixed. Secure operating systems must be able to attach labels to dynamically created subjects and objects and even enable label transitions.

A *labeling state* assigns labels to new subjects and objects. Figure 2.2 shows that processes and files are associated with labels in a fixed protection state. When `newfile` is created, it must be assigned one of the object labels in the protection state. In Figure 2.2, it is assigned the `secret` label. Likewise, the process `newproc` is also labeled as `unclassified`. Since the access matrix does not permit `unclassified` subjects with access to `secret` objects, `newproc` cannot access `newfile`. As for the protection state, in a secure operating system, the labeling state must be defined by trusted administrators and immutable during system execution.

A *transition state* enables a secure operating system to change the label of a process or a system resource. For a process, a label transition changes the permissions available to the process (i.e., its protection domain), so such transitions are called *protection domain transitions* for processes. As an example where a protection domain transition may be necessary, consider when a process executes a different program. When a process performs an `execve` system call the process image (i.e., code and data) of the program is replaced with that of the file being executed. Since a different program is run as a result of the `execve` system call, the label associated with that process may need to be changed as well to indicate the requisite permissions or trust in the new image.

A transition state may also change the label of a system resource. A label transition for a file (i.e., object or resource) changes the accessibility of the file to protection domains. For example, consider the file `acct` that is labeled `trusted` in Figure 2.2. If this file is modified by a process with an `untrusted` label, such as `other`, a transition state may change its label to `untrusted` as well. The Low-Water Mark (LOMAC) policy defines such kind of transitions [101, 27] (see Chapter 5). An alternative would be to change the protection state to prohibit `untrusted` processes from modifying `trusted` files, which is the case for other policies. As for the protection state and labeling state, in a secure operating system, the transition state must be defined by trusted administrators and immutable during system execution.

---

[2]Historically, the term *mandatory access control* has been used to define a particular family of access control models, lattice-based access control models [271]. Our use of the terms *mandatory protection system* and *mandatory access control system* are meant to include historical MAC models, but our definition aims to be more general. We intend that these terms imply models whose sets of labels are immutable, including these MAC models and others, which are administered only by trusted subjects, including trusted software and administrators. We discuss the types of access control models that have been used in MAC systems in Chapter 5.

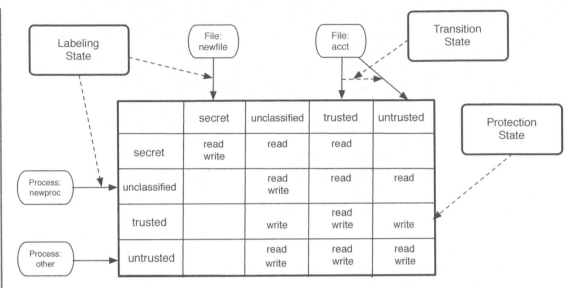

Figure 2.2: A Mandatory Protection System: The *protection state* is defined in terms of labels and is immutable. The immutable *labeling state* and *transition state* enable the definition and management of labels for system subjects and objects.

2.2 REFERENCE MONITOR

A *reference monitor* is the classical access enforcement mechanism [11]. Figure 2.3 presents a generalized view of a reference monitor. It takes a request as input, and returns a binary response indicating whether the request is *authorized* by the reference monitor's access control policy. We identify three distinct components of a reference monitor: (1) its interface; (2) its authorization module; and (3) its policy store. The interface defines where the authorization module needs to be invoked to perform an authorization query to the protection state, a labeling query to the labeling state, or a transition query to the transition state. The authorization module determines the exact queries that are to be made to the policy store. The policy store responds to authorization, labeling, and transition queries based on the protection system that it maintains.

Reference Monitor Interface   The reference monitor interface defines where protection system queries are made to the reference monitor. In particular, it ensures that all security-sensitive operations are authorized by the access enforcement mechanism. By a *security-sensitive operation*, we mean an *operation* on a particular *object* (e.g., file, socket, etc.) whose execution may violate the system's security requirements. For example, an operating system implements file access operations that would allow one user to read another's secret data (e.g., private key) if not controlled by the operating system. Labeling and transitions may be executed for authorized operations.

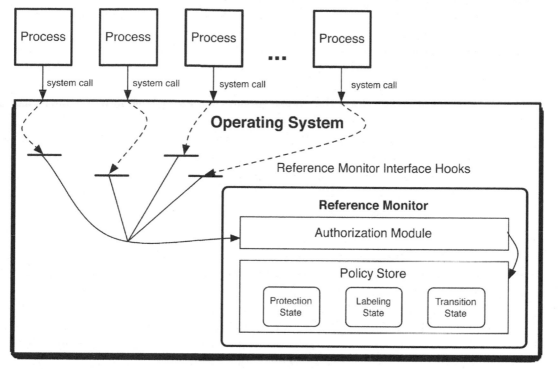

Figure 2.3: A *reference monitor* is a component that authorizes access requests at the *reference monitor interface* defined by individual *hooks* that invoke the reference monitor's *authorization module* to submit an authorization query to the *policy store*. The policy store answers authorization queries, labeling queries, and label transition queries using the corresponding states.

The reference monitor interface determines where access enforcement is necessary and the information that the reference monitor needs to authorize that request. In a traditional UNIX file open request, the calling process passes a file path and a set of operations. The reference monitor interface must determine what to authorize (e.g., directory searches, link traversals, and finally the operations for the target file's inode), where to perform such authorizations (e.g., authorize a directory search for each directory inode in the file path), and what information to pass to the reference monitor to authorize the open (e.g., an inode reference). Incorrect interface design may allow an unauthorized process to gain access to a file.

Authorization Module   The core of the reference monitor is its authorization module. The authorization module takes interface's inputs (e.g., process identity, object references, and system call name), and converts these to a query for the reference monitor's policy store. The challenge for the authorization module is to map the process identity to a subject label, the object references to an object label, and determine the actual operations to authorize (e.g., there may be multiple operations per interface). The protection system determines the choices of labels and operations, but the authorization module must develop a means for performing the mapping to execute the "right" query.

For the open request above, the module responds to the individual authorization requests from the interface separately. For example, when a directory in the file path is requested, the authorization module builds an authorization query. The module must obtain the label of the subject responsible for the request (i.e., requesting process), the label of the specified directory object (i.e., the directory inode), and the protection state operations implied the request (e.g., read or search the directory). In some cases, if the request is authorized by the policy store, the module may make subsequent requests to the policy store for labeling (i.e., if a new object were created) or label transitions.

Policy Store   The policy store is a database for the protection state, labeling state, and transition state. An authorization query from the authorization module is answered by the policy store. These queries are of the form {subject_label, object_label, operation_set} and return a binary authorization reply. Labeling queries are of the form {subject_label, resource} where the combination of the subject and, optionally, some system resource attributes determine the resultant resource label returned by the query. For transitions, queries include the {subject_label, object_label, operation, resource}, where the policy store determines the resultant label of the resource. The resource may be either be an active entity (e.g., a process) or a passive object (e.g., a file). Some systems also execute queries to authorize transitions as well.

2.3   SECURE OPERATING SYSTEM DEFINITION

We define a *secure operating system* as a system with a reference monitor access enforcement mechanism that satisfies the requirements below when it enforces a mandatory protection system.

Definition 2.5.   A *secure operating system* is an operating system where its access enforcement satisfies the *reference monitor concept* [11].

Definition 2.6.   The *reference monitor concept* defines the necessary and sufficient properties of any system that securely enforces a mandatory protection system, consisting of three guarantees:

1. **Complete Mediation**: The system ensures that its access enforcement mechanism mediates all security-sensitive operations.

2. **Tamperproof**: The system ensures that its access enforcement mechanism, including its protection system, cannot be modified by untrusted processes.

3. **Verifiable**: The access enforcement mechanism, including its protection system, "must be small enough to be subject to analysis and tests, the completeness of which can be assured" [11]. That is, we must be able to prove that the system enforces its security goals correctly.

The reference monitor concept defines the *necessary* and *sufficient* requirements for access control in a secure operating system [145]. First, a secure operating system must provide complete mediation of all security-sensitive operations. If all these operations are not mediated, then a security requirement may not be enforced (i.e., a secret may be leaked or trusted data may be modified by an untrusted process). Second, the reference monitor system, which includes its implementation and the protection system, must all be tamperproof. Otherwise, an attacker could modify the enforcement function of the system, again circumventing its security. Finally, the reference monitor system, which includes its implementation and the protection system, must be small enough to verify the correct enforcement of system security goals. Otherwise, there may be errors in the implementation or the security policies that may result in vulnerabilities.

A challenge for the designer of secure operating system is how to precisely achieve these requirements.

Complete Mediation   Complete mediation of security-sensitive operations requires that all program paths that lead to a security-sensitive operation be mediated by the reference monitor interface. The trivial approach is to mediate all system calls, as these are the entry points from user-level processes. While this would indeed mediate all operations, it is often insufficient. For example, some system calls implement multiple distinct operations. The open system call involves opening a set of directory objects, and perhaps file links, before reaching the target file. The subject may have different permission for each of these objects, so several, different authorization queries would be necessary. Also, the directory, link, and file objects are not available at the system call interface, so the interface would have to compute them, which would result in redundant processing (i.e., since the operating system already maps file names to such objects). But worst of all, the mapping between the file name passed into an open system call and the directory, link, and file objects may be changed between the start of the system call and the actual open operation (i.e., by a well-timed rename operation). This is called a *time-of-check-to-time-of-use* (TOCTTOU) attack [30], and is inherent to the open system call.

As a result, reference monitors require interfaces that are embedded in the operating system itself in order to enforce complete mediation correctly. For example, the Linux Security Modules (LSM) framework [342] (see Chapter 9), which defines the mediation interface for reference monitors in Linux does not authorize the open system call, but rather each individual directory, link, and file open after the system object reference (i.e., the inode) has been retrieved. For LSM, tools have been built to find bugs in the complete mediation demanded of the interface [351, 149], but it is difficult to verify that a reference monitor interface is correct.

Tamperproof   Verifying that a reference monitor is tamperproof requires verifying that all the reference monitor components, the reference monitor interface, authorization module, and policy store, cannot be modified by processes outside the system's *trusted computing base* (TCB) (see Chapter 1). This also implies that the TCB itself is high integrity, so we ultimately must verify that the entire TCB cannot be modified by processes outside the TCB. Thus, we must identify all the ways that the TCB can be modified, and verify that no untrusted processes (i.e., those outside the TCB) can perform such modifications. First, this involves verifying that the TCB binaries and data files are unmodified. This can be accomplished by a multiple means, such as file system protections and binary verification programs. Note that the verification programs themselves (e.g., Tripwire [169]) must also be protected. Second, the running TCB processes must be protected from modification by untrusted processes. Again, system access control policy may ensure that untrusted processes cannot communicate with TCB processes, but for TCB processes that may accept inputs from untrusted processes, they must protect themselves from malicious inputs, such as buffer overflows [232, 318], format string attacks [305], and return-to-libc [337]. While defenses for runtime vulnerabilities are fundamental to building tamperproof code, we do not focus on these software engineering defenses in this book. Some buffer overflow defenses, such as StackGuard [64] and stack randomization [121], are now standard in compilers and operating systems, respectively.

Second, the policy store contains the mandatory protection system which is a MAC system. That is, only trusted administrators are allowed to modify its states. Unfortunately, access control policy is deployment-specific, so administrators often will need to modify these states. While administrators may be trusted they may also use untrusted software (e.g., their favorite editor). The system permissions must ensure that no untrusted software is used to modify the mandatory protection system.

Tamperproofing will add a variety of specific security requirements to the system. These requirements must be included in the verification below.

Verifiable   Finally, we must be able to verify that a reference monitor and its policy really enforce the system security goals. This requires verifying the correctness of the interface, module, and policy store software, and evaluating whether the mandatory protection system truly enforces the intended goals. First, verifying the correctness of software automatically is an unsolved problem. Tools have been developed that enable proofs of correctness for small amounts of code and limited properties (e.g., [18]), but the problem of verifying a large set of correctness properties for large codebases appears intractable. In practice, correctness is evaluated with a combination of formal and manual techniques which adds significant cost and time to development. As a result, few systems have been developed with the aim of proving correctness, and any comprehensive correctness claims are based on some informal analysis (i.e., they have some risk of being wrong).

Second, testing that the mandatory protection system truly enforces the intended security goals appears tractable, but in practice, the complexity of systems makes the task difficult. Because the protection, labeling, and transition states are immutable, the security of these states can be assessed.

For protection states, some policy models, such as Bell-LaPadula [23] and Biba [27], specify security goals directly (see Chapter 5), but these are idealizations of practical systems. In practice, a variety processes are trusted to behave correctly, expanding the TCB yet further, and introducing risk that the security goals cannot be enforced. For operating systems that have fine-grained access control models (i.e., lots of unique subjects and objects), specifying and verifying that the policy enforces the intended security goals is also possible, although the task is significantly more complex.

For the labeling and transition states, we must consider the security impact of the changes that these states enable. For example, any labeling state must ensure that any label associated with a system resource does not enable the leakage of data or the modification of unauthorized data. For example, if a `secret` process is allowed to create `public` objects (i.e., those readable by any process), then data may be leaked. The labeling of some objects, such as data imported from external media, presents risk of incorrect labeling as well.

Likewise, transition states must ensure that the security goals of the system are upheld as processes and resources are relabeled. A challenge is that transition states are designed to enable *privilege escalation*. For example, when a user wants to update their password, they use an unprivileged process (e.g., a shell) to invoke privileged code (e.g., the `passwd` program) to be run with the privileged code's label (e.g., UNIX `root` which provides full system access). However, such transitions may be insecure if the unprivileged process can control the execution of the privileged code. For example, unprivileged processes may be able to control a variety of inputs to privileged programs, including libraries, environment variables, and input arguments. Thus, to verify that the system's security goals are enforced by the protection system, we must examine more than just the protection system's states.

2.4   ASSESSMENT CRITERIA

For each system that we examine, we must specify precisely how each system enforces the reference monitor guarantees in order to determine how an operating system aims to satisfy these guarantees. In doing this, it turns out to be easy to expose an insecure operating system, but it is difficult to define how close to "secure" an operating system is. Based on the analysis of reference monitor guarantees above, we list a set of dimensions that we use to evaluate the extent to which an operating system satisfies these reference monitor guarantees.

1. **Complete Mediation**: How does the reference monitor interface ensure that all security-sensitive operations are mediated correctly?

   In this answer, we describe how the system ensures that the subjects, objects, and operations being mediated are the ones that will be used in the security-sensitive operation. This can be a problem for some approaches (e.g., *system call interposition* [3, 6, 44, 84, 102, 115, 171, 250]), in which the reference monitor does not have access to the objects used by the operating system. In some of these cases, a race condition may enable an attacker to cause a different object to be accessed than the one authorized by reference monitor [30].

2. **Complete Mediation**: Does the reference monitor interface mediate security-sensitive operations on all system resources?

   We describe how the mediation interface described above mediates all security-sensitive operations.

3. **Complete Mediation**: How do we verify that the reference monitor interface provides complete mediation?

   We describe any formal means for verifying the complete mediation described above.

4. **Tamperproof**: How does the system protect the reference monitor, including its protection system, from modification?

   In modern systems, the reference monitor and its protection system are protected by the operating system in which they run. The operating system must ensure that the reference monitor cannot be modified and the protection state can only be modified by trusted computing base processes.

5. **Tamperproof**: Does the system's protection system protect the trusted computing base programs?

   The reference monitor's tamperproofing depends on the integrity of the entire trusted computing base, so we examine how the trusted computing base is defined and protected.

6. **Verifiable**: What is basis for the correctness of the system's trusted computing base?

   We outline the approach that is used to justify the correctness of the implementation of all trusted computing base code.

7. **Verifiable**: Does the protection system enforce the system's security goals?

   Finally, we examine how the system's policy correctly justifies the enforcement of the system's security goals. The security goals should be based on the models in Chapter 5, such that it is possible to test the access control policy formally.

While this is undoubtedly an incomplete list of questions to assess the security of a system, we aim to provide some insight into why some operating systems cannot be secure and provide some means to compare "secure" operating systems, even ones built via different approaches.

We briefly list some alternative approaches for further examination. An alternative definition for penetration-resistant systems by Gupta and Gligor [122, 123] requires tamperproofing and complete mediation, but defines "simple enough to verify" in terms of: (1) consistency of system global variables and objects; (2) timing consistency of condition checks; and (3) elimination of undesirable system/user dependencies. We consider such goals in the definition of the tamperproofing requirements (particularly, number three) and the security goals that we aim to verify, although we do not assess the impact of timing in this book in detail. Also, there has been a significant amount of work on formal verification tools as applied to formal specifications of security for assessing the information flows among system states [79, 172, 331]. For example, Ina Test and Ina Go are symbolic execution tools that interpret the formal specifications of a system and its initial conditions, and compare the resultant states to the expected conditions of those states. As the formal specification of systems and expectations are complex, such tools have not achieved mainstream usage, but remains an area of exploration for determining practical methods for verifying systems (e.g., [134, 132]).

2.5 SUMMARY

In this chapter, we define the fundamental terminology that we will use in this book to describe the secure operating system requirements.

First, the concept of a *protection system* defines the system component that enforces the access control in an operating system. A protection system consists of a *protection state* which describes the operations that are permitted in a system and *protection state operations* which describe how the protection state may be changed. From this, we can determine the operations that individual processes can perform.

Second, we identify that today's commercial operating systems use protection systems that fail to truly enforce security goals. We define a *mandatory protection system* which will enforce security in the face of attacks.

Third, we outline the architecture of an access enforcement mechanism that would be implemented by a protection system. Such enforcement mechanisms can enforce a mandatory protection state correctly if they satisfy the guarantees required of the *reference monitor concept*.

Finally, we define requirements for a secure operating system based on a reference monitor and mandatory protection system. We then describe how we aim to evaluate the operating systems described in this book against those secure operating system requirements.

Such a mandatory protection system and reference monitor within a mediating, tamperproof, and verifiable TCB constitute the *trust model* of a system, as described in Chapter 1. This trust model provides the basis for the enforcement of system security goals. Such a trust model addresses the system *threat model* based on achievement of the reference monitor concept. Because the reference monitor mediates all security-sensitive operations, it and its mandatory protection state are tamperproof, and both are verified to enforce system security goals, then is it possible to have a comprehensive *security model* that enforces a system's security goals.

Although today's commercial operating systems fail to achieve these requirements in many ways, a variety of operating systems designs in the past and that are currently available are working toward meeting these requirements. Because of the interest in secure operating systems and the variety of efforts being undertaken, it is more important than ever to determine how an operating system design aims to achieve these requirements and whether the design approaches will actually satisfy these secure operating system requirements.

CHAPTER 3

Multics

In this chapter, we examine the first modern operating system, the Multics system [62]. Multics was a large, long-term operating system project where many of our fundamental operating systems concepts, including those for secure operating systems, were invented. For operating systems, these concepts include segmented and virtual memory, shared memory multiprocessors, hierarchical file systems, and online reconfiguration among others. For secure operating systems, the ideas of the reference monitor, protection systems, protection domain transitions, and multilevel security policies (see Chapter 5), among others, were also pioneered in Multics.

We review the history of the Multics operating system, its security features, and then evaluate how Multics satisfies the requirements of a secure operating system developed in Chapter 2. The Multics history has a number of interesting twists and turns that are relevant to those considering the development or use of secure operating systems today. From a technology perspective, the Multics system implementation demonstrates, often more extensively than many of the other secure operating systems that we will study, the secure operating system definition in Chapter 2. In subsequent chapters, we will often compare other operating system implementations to Multics.

3.1 MULTICS HISTORY

The Multics project began in 1963 with the aim of building a comprehensive, timesharing operating system [217, 126]. Multics emerged from the work of an early timesharing system called the Compatible Timesharing System (CTSS) [87], a project led by Fernando Corbató at MIT. Until 1960 or so, computers all ran batch systems, where individual jobs were run in sequence, one at a time. Often users would have to wait hours or days to receive the results of their program's executions. Early timesharing systems could support a small number of programs by swapping the contents of those not running to tape (i.e., there was not enough memory to keep them in memory). Operating systems of the day, whether batch or timesharing, supported a very different set of functions as well, aiming at building programs, loading them into memory, and automating some basic tasks of the system administrators.

CTSS was demonstrated in 1961, and its success motivated the Advanced Research Projects Agency (ARPA) of the US Department of Defense (DoD) to create Project MAC, which stood for Multi-Access Computer among other things [1]. Project MAC proposed to build general-purpose, timesharing services to support large numbers of users simultaneously. It would have to support functions that would enable the multiplexing of devices for multiple processes (i.e., running programs), scheduling of those processes, communication between processes, and protection among processes. Based on a summer study in 1962, the specifications for Multics were developed and submitted for

---

[1]"MAC stood for Multiple Access Computers on the 5th floor of 545 Tech Square and Man and Computer on the 9th floor [217]."

bid in 1963. Folklore has it that IBM was not interested in Project MAC's ideas for paging and segmentation, so instead General Electric (GE) was chosen to build the hardware for the project, the eventual GE 645. Bell Labs joined the software development in 1965.

The Multics project had very ambitious and revolutionary goals, so it is not surprising that the project had its moments of intrigue. The project called for delivery of the project in two and a half years, but delivery of the GE 645 hardware was delayed such that Multics was not self-hosting [2] until 1968. ARPA considered terminating the project in 1969, and Bell Labs dropped out of the project in 1969 as well. Ultimately, the Multics system itself proved to be bigger, slower, and less reliable than expected, but a number of significant operating systems and security features were developed that live on in modern systems, such as the UNIX system that was developed by some of the associated Bell Labs researchers after they left the Multics project, see Chapter 4.

Multics emerged as a commercial product in 1973. Honeywell purchased GE's hardware business in 1970, and Honeywell sold Multics systems until 1985. As Multics systems were expensive ($7 million initially), only about 80 licenses were sold. The primary purchasers were government organizations (e.g., US Air Force), university systems (e.g., the University of Southwest Louisiana and the French university system), and large US corporations (e.g., General Motors and Ford). Multics development was terminated in 1985, but Multics systems remained in use until 2000. The last running Multics system was used by the Canadian Department of National Defense.

The Multics project was unusual for its breadth of tasks, diversity of partners, and duration under development. While any hardware project requires the development of an ecosystem (e.g., compilers, system software, etc.) for people to program to the hardware, the Multics project was both a substantial hardware project, a revolutionary operating systems project, and a groundbreaking security project. This breadth of tasks would be daunting today. Secondly, the Multics project team represented university and industry researchers in addition to a variety of government and industry engineers. Many members of the project were among our greatest computer minds, so it is not easy to assemble such a group. Thirdly, an astounding thing is that the project persisted for nearly 10 years before any commercial product was released. In today's competitive environment, such a long pre-production phase would be highly unusual. As a result, the Multics project had a unique situation that enabled them to pursue ambitious, long-term goals, such as building a secure operating system.

3.2   THE MULTICS SYSTEM

The Multics system architecture is a layered architecture where executing programs may be permitted to access named system resources that are organized hierarchically. In this section, we first examine the basic principles of the system, then its security features. This information is culled from the many research documents published on the Multics system. The most comprehensive documents written about Multics were Organick's book [237] and the Final Report of the project [280].

---

[2] A *self-hosting* system can be used to develop new versions of itself.

3.2.1   MULTICS FUNDAMENTALS

The fundamental concepts in the Multics system are processes and segments. *Processes* are the executable contexts in Multics—that is, they run program code. All code, data, I/O devices, etc. that may be accessed by a process are stored as *segments*. Segments are organized into a hierarchy of directories that may contain directories or segments.

A process's *protection domain* defines the segments that it can access. A Multics process's *protection domain* consists of the segments that could be loaded into its descriptor segment and the operations that the process could then perform on those segments. Each segment is associated with its accessibility—i.e., the *subjects* whose processes can access the segment and the operations that they are allowed to perform. Multics has three different ways of expressing accessibility that we will describe in Section 3.2.1.

Segments are addressable either locally within the process's context or by name from secondary storage (i.e., analogous to modern file systems). For segments already in a process's context, Figure 3.1 shows that each process is associated with its own *descriptor segment* that contains a set of *segment descriptor words* (SDWs) that refer to all the segments that the process can directly access. That is, these segments are directly addressable by the process in the system's memory [3].

When Multics process requests a segment that is not already in its descriptor segment, it must name the segment using what is analogous to a file path. Like modern file systems, Multics segments are named hierarchically. For example, the name /U2/War/NewYearsDay is processed starting with the root directory, continuing with subsequent descendant directories (i.e., U2 and War), and finishing with the name of the actual segment (e.g., the NewYearsDay segment). Thus, Multics segment access provided a blueprint for later hierarchical file systems of UNIX and beyond. If the process's subject has the permissions to perform the requested operation on the segment, then a new SDW is created with those permissions and is loaded into the process's descriptor segment. Note that the process must also have the access to all the directories in the segment's path as well to access the segment.

3.2.2   MULTICS SECURITY FUNDAMENTALS

Multics security depends on some fundamental concepts that we introduce before we detail the protection system and reference monitor. These concepts include the Multics *supervisor*, *protection rings*, and Multics *segment descriptor words*.

Figure 3.2 shows the actions that take place when a user logs into a Multics system. First, a user login requires that a component of the trusted computing base (TCB) verify the user's password and build a process for the user to perform their processing. User logins are implemented by a process called the *answering service*. To authenticate the user, the answering service must retrieve the password segment from the file system by loading the password SDW into its descriptor segment. The loading and subsequent use of the password segment must be authorized by the core Multics

---

[3] Of course, a segment may have been swapped out to secondary storage, but from the point of view of the process, the segment is available in memory. It will be swapped in invisibly by the Multics kernel.

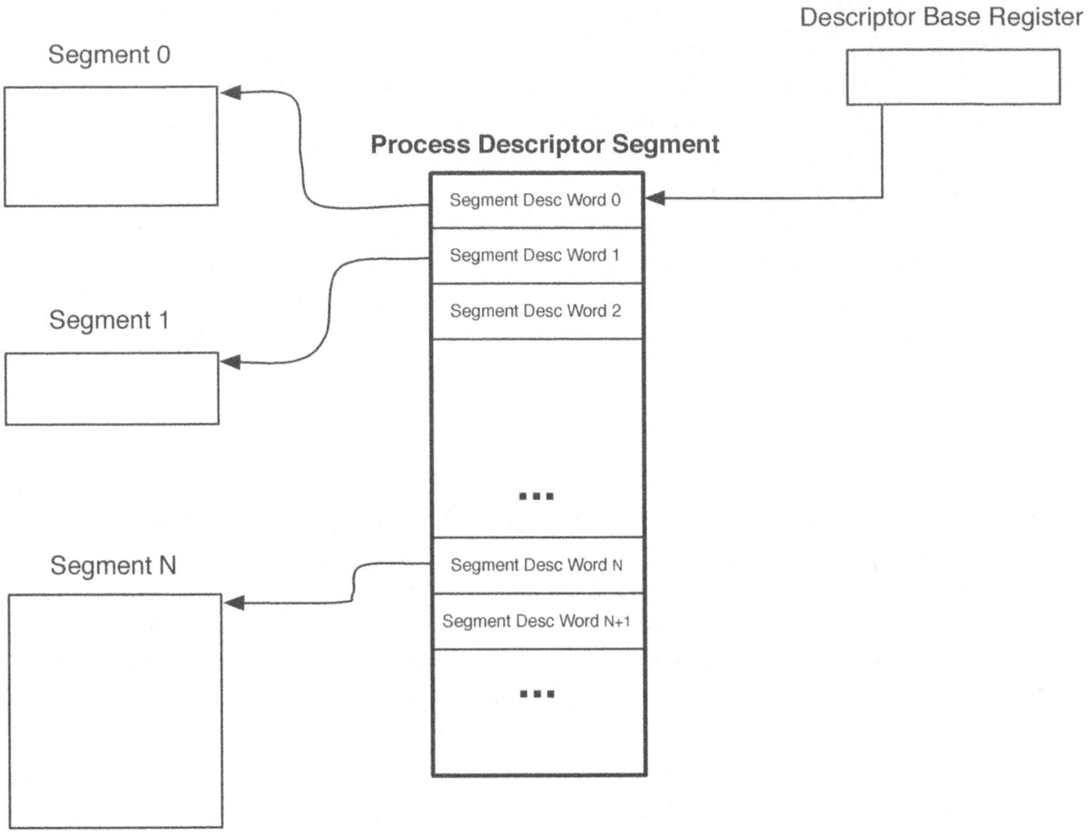

Figure 3.1: Multics process's segment addressing (adapted from [159]). Each process stores a reference to its *descriptor segment* in its *descriptor base register*. The descriptor segment stores *segment descriptor words* that reference each of the process's active segments (e.g., segments 0, 1, and *N*).

component, the *supervisor* [322]. If authorized, a SDW for the password segment is loaded into the answering service's descriptor segment. The supervisor implements the most trusted functionality in the Multics system, such as authorization, segmentation, file systems, I/O, scheduling, etc. Early Multics systems also included dynamic linking functionality in the supervisor, but that was later removed [62] and is also implemented in user-space in modern systems.

The supervisor is isolated from other processes by *protection rings* [281]. Protection rings form a hierarchical layering from the most privileged ring, ring 0 where the most-privilege code in the supervisor runs, to the least privileged ring. There were 64 rings in the GE 645 Multics system, but only 8 were implemented in GE 645 hardware and the rest by some software tricks. The supervisor

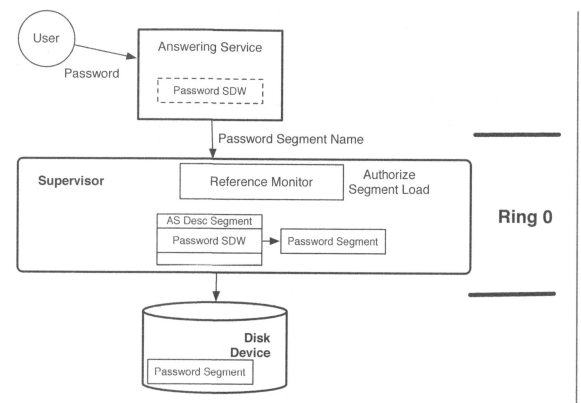

Figure 3.2: The Multics login process. The user's password is submitted to the Multics *answering service* which must check the password against the entries in the *password segment*. The Multics *supervisor* in the privileged *protection ring 0* authorizes access to this segment and adds a SDW for it to the answering service's descriptor segment. The answering service cannot modify its own descriptor segment.

is protected from other processes because only its segments are assigned to rings 0 and 1 [4], and no process running in a higher ring can modify its segments. Thus, processes can only cause a modification of the supervisor's state by invoking supervisor code that runs in ring 0. Multics defines mechanisms to protect the supervisor from malicious input in these calls. The Multics design aimed for layering of function as advocated by other systems of the time, such as the THE system [76], but the rings were ultimately used as a simple, coarse-grained mechanism to protect the integrity of the supervisor and other trusted processes from untrusted code. Of course, modern processors also protect their operating systems using protection rings, although only two levels, supervisor and user, are typically utilized.

---

[4]The Multics supervisor is divided into ring 0 components, including access control, I/O, and memory management, and ring 1 components that are less primitive, such as accounting, stream management, and file system search.

If the user and password match, then the answering service creates a user process with the appropriate code and data segments for running on behalf of that user. Each live process segment is accessed via a *segment descriptor word* (SDW) as mentioned above. Figure 3.3 shows the SDW layout [281]. The SDW contains the address of the segment in memory, its length, its *ring brackets*,

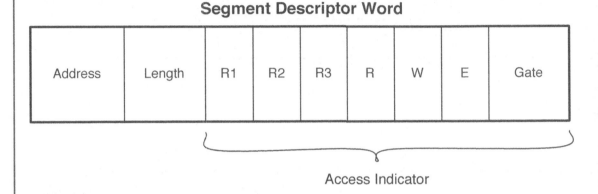

Figure 3.3: Structure of the Multics *segment descriptor word* (SDW): in addition to the segment's address and length, the SDW contains access indicators including *ring brackets* (i.e., *R1, R2, R3*), the process's ACL for the segment (i.e., the *rwe* bits), and the *number of gates* for the segment.

its process's permissions (rwe) for this segment, and, for code segments, the number of gates defined for the segment. When the process references an SDW, its ring bracket limits access based on the current ring in which the process is running. The process permissions (rwe) limit the operations that the process can ever perform on this segment. We examine the meaning of the SDW access fields below.

3.2.3   MULTICS PROTECTION SYSTEM MODELS

The Multics protection system consists of three different, interacting models that individually provide distinct aspects of the overall Multics protection system. For simplicity, We introduce the models in isolation first, and the describe the overall authorization process.

Access Control List   First, each object (i.e., segment or directory) is associated with its own *access control list* (ACL). Each ACL entry specifies a user identity of processes and the operations that processes with that identity can perform on this object. Note that a user may be specified using wildcards to represent groups of users. Segments and directories have different operation sets. Segments may be read (r), written (w), or executed (e), and directories may be accessed to obtain the status of the entry (s), modify an entry (i.e., delete or modify ACLs, (m), or append an entry to the directory (a). Note that the ACLs for a segment are stored in its parent directory, so access is checked at the parent. Also, any modification of an ACL for a segment requires the modification permission on the parent directory.

Example 3.1.   Examples of ACLs on a segment include:

```
rew    Jaeger.SysAdmin.*
r      Backup.SysDaemon.*
rw     *.SysAdmin.*
```

Also, examples of directory ACLs include:

```
sma    Jaeger.SysAdmin.*
s      Backup.SysDaemon.*
sm     *.SysAdmin.*
```

When a process requests access to a segment, the ACL of the segment is checked to determine if the user associated with the process has an entry in the ACL that permits the requested operations. If so, the reference monitor authorizes the construction of an SDW with those operations.

Rings and Brackets   Multics also limits access based on the protection ring of the process. Each segment is associated with a *ring bracket* specification that defines read, write, and execute permissions of processes over that segment. Also, protection domain transition rules are defined by these brackets. First, a segment's *access bracket* defines the ranges of rings that can read and write to a segment. An access bracket is specified by a range of rings $(r1, r2)$ where $r1 \leq r2$ (i.e., $r1$ is more privileged than $r2$). Suppose a process is running in ring r, then the access rights of that process to a segment with an access bracket of $(r1, r2)$ are determined by:

- If $r < r1$, then the process can read and write to the segment.

- If $r1 \leq r \leq r2$, then the process can read the segment only.

- If $r2 < r$, then the process has no access to the segment.

Such a policy ensures that lower rings (i.e., more privileged) have strictly greater access to segments than the higher rings.

Multics also uses rings to control the invocation of code segments. A second access specification, the *call bracket*, is used along with the access bracket to determine how a process in ring r invokes a code segment. The call bracket is $(r2, r3)$, where $r2$ is same $r2$ as in the access bracket and $r2 \leq r3$. If a process at ring r tries to invoke a code segment with an access bracket of $(r1, r2)$ and a call bracket of $(r2, r3)$, the following cases are possible:

- If $r < r1$, then the process can execute the code segment, but there is a ring transition from r to a lower privileged ring $r1 \leq r' \leq r2$ specified by the segment (typically, $r1 == r2$, so the transition is obvious).

- If $r1 \leq r \leq r2$, then the process invokes the code segment in its current ring r (i.e., no ring transition).

- If $r2 \leq r \leq r3$, then the process can execute the code segment, there is a ring transition from r to the higher privileged ring r' if authorized by the *gates* in the code segment's SDW.

- If $r3 < r$, then the process cannot invoke the code segment.

The call brackets not only define execute privilege based on the process's current protection ring, but they also define transition rules describing the requirements for protection domain transition (e.g., if authorized by all gates) and the resultant ring number for the executing code. Call brackets are the only means of describing transition state in the Multics system.

Multilevel Security   Multics pioneered the enforcement of *Multilevel Security* [23, 326] (MLS) in operating systems [5]. An MLS policy prevents a subject from reading data that is more secret than the subject or writing data to less secret objects. A detailed description of MLS and its semantics is provided in Chapter 5.

In Multics, each directory stores a mapping from each segment to a secrecy level. Also, Multics stores an association between each process and its secrecy level. A request is authorized if one of three conditions are met:

1. **Write**: The process requests write access only and the level of the segment/directory is greater than (i.e., *dominates*) or equal to the level of the process.

2. **Read**: The process requests read access only and the level of the segment/directory is less than (i.e., *dominated by*) or equal to the level of the process.

3. **Read/Write**: The process requests read and write access and the level of the segment/directory is the same as the process or the process is designated as trusted.

Intuitively, we can see that a process can only read a segment/directory if its level is more secret or the same as the level of the object and write a segment/directory if its level is less secret or the same as that of the object. This prevents information leakage by preventing a process from reading information that is more secret than its secrecy level and preventing a process from writing its information to objects of a lower secrecy level. In Chapter 5, we formally defines MLS secrecy enforcement.

3.2.4   MULTICS PROTECTION SYSTEM

Multics's protection system consists of these three policies. When a segment is requested, all three policies must authorize the request for it to be allowed. If the requested operation is a *read*, the ACL

---

[5] MLS is called the Access Isolation Mechanism (AIM) in Multics documentation. We will use the current term of MLS for such access control systems.

is checked to determine if the user has access, the MLS policy is checked to verify that the object's secrecy level is dominated by or equal to the process's, and the access bracket is checked to determine whether the process has read access to the object's segment ($r \leq r2$). When the requested operation is a *write*, the ACL is checked for write access, the MLS policy is checked to verify that the object's secrecy level dominates or is equal to the process's, and the access bracket must permit the current ring write access ($r < r1$).

An *execute* request is handled similarly, except the call bracket is used instead of the access bracket, and the request may result in a protection domain transition. The process must have execute permission in the segment's ACL, the MLS policy must permit reading the segment, and the call bracket must permit execution.

Execution of a segment may also result in a transition from the process's current ring r to the ring specified by the segment (we call this r') based on the call bracket. There are two cases. First, when this process invokes a code segment with a call bracket where $r < r1$, then the process must transition to r' (i.e., a lower-privileged ring). Second, when this process invokes a code segment with a call bracket where $r2 \leq r \leq r3$, then the process must use one of the valid segment gates as an entry point and transition to r' (i.e., enter a higher-privileged ring if the gates allow).

As described in Chapter 2, a secure protection system consists of a protection state, a labeling state, and a transition state that may only be administered by trusted subjects. Multics defines its protection state based on these three models. The ring brackets define the allowed protection domain transitions in the system. There are no object transitions specified in the Multics policy. Labeling is not specifically defined in the Multics policy. Presumably, new segments are assigned the MLS labels and ring brackets from their creator, but this is not specified.

Both the ACL and ring bracket policies are *discretionary access control policies*. That is, the ACLs and ring brackets for a segment may modified by any process that has the modify privilege to the segment's parent directory. Only the MLS policy is *nondiscretionary* or *mandatory*. The MLS policy is loaded with the system at boot-time and is otherwise immutable.

3.2.5 MULTICS REFERENCE MONITOR
The Multics reference monitor is implemented by the supervisor. Each Multics instruction either accesses a segment via a directory or via a SDW, so authorization is performed on each instruction. Originally, the supervisor performed such authorizations, but eventually hardware extensions enabled most SDW authorizations to be performed directly by the hardware [281], as we now are accustomed. The supervisor then became responsible for setting up the process's descriptor segment and preventing the process from modifying it.

In addition to protection state queries, the supervisor also performs protection domain transitions by changing the process's ring as described above. Accessing a code segment has three allowed cases, two that result in a ring transition. Invoking code in a ring below (i.e., more privileged than) the access bracket results in a ring transition to a more-privileged ring. Such transitions require entry through a special gate segment that verifies: (1) the number of arguments expected; (2) the data type on each argument; and (3) access requirements for each argument (e.g., read only or read-write). The gate segment, also called a *gatekeeper*, aims to protect the invoked code from potentially malicious input from lower-privileged code. The called procedure must also not depend on the caller for stack memory, and it must return to the calling code in the proper ring number r.

The transition to a lower-privileged ring also generates some security issues. In this case, we may leak information as a result of the call to a lower-privileged ring and that the higher privileged code must protect itself on a return. In the first case, we need to ensure that the called procedure in a high ring (i.e., less-privileged ring) has access to the procedure arguments. Since the granularity of control is a segment, each segment in which an argument is contained must be accessible to the called procedure. Multics can enforce protection on segments, such that the called procedure does not get unauthorized access, but that may result in program failures. Thus, some form of copying is necessary. For example, the supervisor copies arguments from its segment to another segment accessible to the called procedure. However, the caller must be careful not to copy unauthorized information, such as private keys, that the less-privileged code may be able to use to impersonate the higher-privileged code.

In the second case, Multics enables the caller to provide a gate for the return, called a *return gate*. This mechanism is similar in concept to a call gate, except multiple calls may result in a stack of return gates. Thus, the SDW is unsuitable for return gates. The supervisor must maintain the stack of return gates for the process.

While supervisor functions are implemented in rings 0 and 1, the fundamental reference monitor services are all in ring 0. For example, the file system search utility has been moved to ring 1, such that the determination of a directory or segment from a name is performed there, but authorization of whether this access is permitted is done in the ring 0 supervisor [279]. That is, the code in ring 1 running due to a user's process, may not have an ACL that permits it access to the segment. Thus, ring 0 can limit the actions of code in ring 1. Decisions about what code belongs in ring 0 and ring 1 was an ongoing process throughout the Multics project. Modern operating systems have generally not made such fine-grained distinctions, potentially to their detriment for security. Nonetheless, programming is much simpler in the modern case.

Some services running in less-privileged rings also must be trusted by the supervisor for some functions. For example, the answering service (see Section 3.2.2) performs authentication, so it assigns the user of a process. Clearly, if it is malicious, the process could get unauthorized permissions by being assigned to the wrong user. Also, the administrator must be entrusted with several operations, supported by code that must then be trusted, such as measuring storage usage, performing backups, and changing permission assignments [264]. A TCB was defined for Multics' B2 evaluation (see Chapter 12 for a discussion on system security evaluation), but the Multics architecture continually evolved, such that its TCB evolved over time. In 1973, Saltzer stated that 15% of Multics programs ran in ring 0 [264], so these programs plus administrative and authentication programs minimally defined the Multics TCB. The Multics team recognized that this was a large number of trusted programs, but the resolution of what should be in or what should be out of the TCB remained an ongoing issue until the end of the project.

3.3   MULTICS SECURITY

We evaluate the security of Multics system using the reference monitor principles stated in Chapter 2: complete mediation, tamperproofing, and verifiability. Unlike the commercial operating systems discussed in the next chapter, Multics performs well on these metrics. Nonetheless, we will see that it is difficult to completely achieve these requirements. In the next section, we will discuss how the implementation may cause breaches in security, even in well designed systems.

1. **Complete Mediation**: How does the reference monitor interface ensure that all security-sensitive operations are mediated correctly?

   Since Multics requires that each instruction accesses a segment and each segment access is mediated, Multics provides complete mediation at the segment level. Thus, all security requirements that can be effectively expressed in segments can be mediated in Multics.

   MLS labels for segments are stored in their directories rather than directly in the segments, so Multics must ensure that the mapping between segments and their access classes is used correctly. That is, Multics must prevent a TOCTTOU attack [30] where the attacker can switch the segment assigned a particular name after the access class assigned to the name has been authorized. Traditionally, this is done by restricting a directory to contain only segments of a single access class.

2. **Complete Mediation**: Does the reference monitor interface mediate security-sensitive operations on all system resources?

   Since Multics mediates each segment access at the instruction level, Multics mediates memory access completely. Multics also mediates ring transitions, in both directions. Thus, the reference monitor provides mediation at memory and ring levels. Multics' ring transitions provide argument validation via *gatekeepers*, which is not part of the reference monitor in modern systems (although argument validation is performed procedurally in modern operating system). In practice, the Multics *master mode* permits code to run in a higher ring level without the full ring transition, see Section 3.4 below.

   Also, TCB servers may have finer-grained access control (i.e., within segments), but this is beyond the ability of Multics. If Multics had a server that is trusted to support clients of multiple secrecy levels, it must also ensure that there is no way that an unauthorized information leak can occur (e.g., the *confused deputy problem*, see Chapter 10). In general, such servers must be trusted with such permissions.

3. **Complete Mediation**: How do we verify that the reference monitor interface provides complete mediation?

To verify complete mediation, we need to verify that ring transitions and segment accesses are mediated correctly. These operations are well-defined, so it is straightforward to determine that mediation occurs at these operations. However, the complexity of these operations still made verification difficult. The complexity of addressing resulted in some mediation errors in segment mediation, see Section 3.4.

4. **Tamperproof**: How does the system protect the reference monitor, including its protection system, from modification?

   The Multics reference monitor is implemented by ring 0 procedures. The ring 0 procedures are protected by a combination of the protection ring isolations and system-defined ring bracket policy. The ring bracket policy prevents processes outside of ring 0 from reading or writing reference monitor code or state directly.

   Some ring 0 code must respond to calls from untrusted processes (e.g., system calls). The only way that ring 0 can be accessed by an untrusted process is via a gate. As described above, gates check the format of the arguments to higher-privileged, supervisor code to block malicious inputs. Thus, if the gates are correct, then untrusted processes cannot compromise any ring 0 code, thus protecting the supervisor.

   Multics *master mode* code was not designed to be accessed without a ring transition to ring 0, but this restriction was later lifted, resulting in vulnerabilities (see Section 3.4 below). Thus, a secure Multics system must not include the unprivileged use of master mode as Karger and Schell identify.

5. **Tamperproof**: Does the protection system protect all the trusted computing base programs?

   The Multics TCB consists of the supervisor and some system services in rings 1–3. Multics relegates standard user processing to rings 4 and higher, so trusted code would be placed no higher than ring 3. If we assume that all the code segments in rings 0–3 are part of the trusted computing base, then the TCB is large, but can be protected in the same manner as the supervisor in ring 0.

   The integrity of the TCB depends on its system-defined ring bracket policy. However, the ring bracket policy is a discretionary policy. It can be modified by any subject with modify access to a directory containing a TCB code segment. Should any process in the TCB be compromised, it could undo protections at its ring level, thus potentially compromising the entire ring. If more-privileged rings contain any code that depends on trust in a less-privileged ring that is compromised, then the compromise may spread further. Thus, Multics tamper-protection is "securable" as Saltzer stated, but discretionary access control makes its tamperprotection brittle. See Chapter 7 to see why the use of discretionary access control is problematic.

6. **Verifiable**: What is basis for the correctness of the system's trusted computing base?

The implementation of the Multics TCB is too large to be formally verified [279]. The project's goal was to minimize the Multics implementation as much as possible, such that most, if not all, of the TCB can be verified using manual auditing. This goal was not achieved by the completion of the Multics project, and in fact this limitation motivated the subsequent work in security kernels (see Chapter 6). As we will see in the next section, this resulted in some security problems in Multics.

7. **Verifiable**: Does the protection system enforce the system's security goals?

Verifying that the system's security goals are enforced correctly involves ensuring that the policy: (1) protects the secrecy and integrity of the system and user data by the *protection state*; (2) assigns subjects and objects to the policy labels correctly by the *labeling state*; and (3) ensures that all protection domain transitions protect the secrecy and integrity of the system and user data based on the *transition state* defined by the call bracket rules.

First, the protection state ensures MLS secrecy protection is enforced, although the discretionary management of the ring bracket policy limits integrity protection to the system TCB at best, and only if no TCB process can be compromised. The MLS secrecy policy is a mandatory policy of information flow secrecy goals, so the secrecy goals are enforced by the Multics system given a trusted TCB.

Second, verifying the correct labeling of segments and processes is challenging since much of this labeling is specified manually. The Multics policy does not explicitly state how new processes and segments are labeled, although we would expect that the norm is to inherit the labels of the creating process.

Third, the transition state permits low integrity code in a less-privileged ring to transfer control to high integrity code in a more-privileged ring through either gates or return gates based on the call bracket rules. As discussed above, the security of these transitions depends on the correctness of the gates, but most systems do not even have this level of enforcement.

This informal analysis shows that Multics security is largely very good, but risks remain. In this analysis, we describe powerful mediation, expressive tamperproofing, and verifiable secrecy controls and system integrity controls. However, challenges still remain in managing the scope of the TCB, verifying the correctness of system integrity policies, ensuring integrity protection for all processes, labeling processes and segments correctly, and verifying the correctness of all gates. Saltzer identifies nine areas of security risk in Multics as well [264]. In addition to the issues above, Saltzer mentions the need for secure communication between systems, control of physical access to machines, the weakness of user-specified passwords, the complexity of gate protections for the supervisor, the possibility of leaking secrets via reuse of uncleared memory or storage, excessive privileges for administrators, and others. These issues and challenges are not unique to Multics—as we will see, every secure operating system design will fight with these challenges. For a first attempt at building a secure operating system, the Multics project did an admirable job of identifying issues and proposing solutions, but many difficult issues must be addressed. As Saltzer states, Multics was "designed to be securable," not a single secure configuration.

Of course, building any operating system also requires that the designers consider usability, performance, and maintainability in their design. To a large extent, an operating system is supposed to be invisible to applications. While applications have to use the system's interface to obtain service, the interface should just implement the requests, so programs can run as expected. Of course, the addition of security enforcement may cause programs to no longer work as expected. Requests may be denied for security reasons, and applications may not be prepared to handle such failures. Also, security is supposed to be effectively invisible from a performance perspective. This was a significant problem for Multics, especially given the limited computing power of that time. The small number of deployed Multics systems probably also prevented the usability model of Multics from spreading widely enough to become an accepted norm. As the UNIX community grew to numbers that dwarfed the number of Multics administrators, the computing community came to accept the open, but insecure, approach. Finally, operating systems are complex software components, so they undergo a fair amount of evolution. This was particularly true in the case of Multics, but any system maintenance must still preserve the security guarantees offered by the system. As we will see in the next section, this was not always the case.

3.4 MULTICS VULNERABILITY ANALYSIS

In 1974, a couple of Air Force researchers, Paul Karger and Roger Schell, performed a vulnerability analysis on the Multics system [159]. Unfortunately, the Multics system was too complex for them to do any kind of analysis that may prove the security enforcement of the system (i.e., mediation, tamperproofing, or verifiability), but they examined the system looking for implementation flaws. That is, they were, and still are, firm believers in the Multics approach to building a secure operating system, but they found a number of vulnerabilities in the Multics implementation that raised questions about how to build and maintain secure operating systems.

Karger and Schell's vulnerability analysis investigates whether specific hardware, software, and "procedural" (i.e., configuration) vulnerabilities are present in the Multics system. They found vulnerabilities in each area.

First, a hardware vulnerability was found that would permit an execute instruction to bypass access checking using the SDW. That is, complete mediation could be circumvented due to this vulnerability. The details of the vulnerability require a deeper knowledge of Multics addressing than we provide, but the basic problem is that the Honeywell 645 hardware [6] did not check the SDW access if the segment was reached by a specific format of indirect addressing. Thus, access to the segment containing the indirection was checked for access, but not the segment containing the actual address to be executed. It was found that this error was introduced in a field modification made at MIT and later applied to all processors. While this is was simply an erroneous update, the pressures of balancing performance and security, makes such updates likely. Further, the Multics

---

[6]This analysis was done after Honeywell had purchased GE's computer division in 1970.

project had no tools to enable the verification of security impact of such changes, so errors should not be unexpected.

Second, a variety of software vulnerabilities were reported by Karger and Schell. One of the more significant vulnerabilities was an error caused by misuse of a supervisor mode of execution, called *master mode*. Master mode is an execution state that permits any privileged processor instructions to be executed in the current ring. The original Multics design required master mode code to be restricted to ring 0 only [322]. However, this design choice resulted in all faults (i.e., divide-by-zero, page faults, etc.) incurring a ring transition from the user ring to ring 0 where the fault handler was located and then back to the user ring. A proposal to reduce the overhead on the system was to enable execution of some fault handling (e.g., divide-by-zero and access violations) in the user ring. These faults are reported to user programs anyway by a *signaller* module, so the proposal was to run the signaller in user rings. But, the signaller uses some privileged instructions, so it must run in master mode.

Permitting the signaller to run in master mode in a user ring was deemed secure because of the restricted manner in which the signaller must be invoked, but this code was not designed to protect itself from malicious calls. The problem is that the signaller's code expects a register to be loaded with a reference to a section of the signaller's code when it is called. Unfortunately, the signaller does not check that the register value is legitimate, so when the code became addressable in user rings, it became possible for a malicious user program to set the register to an arbitrary location "permitting him to transfer to an arbitrary location while the CPU was still in master mode [159]." Thus, a significant vulnerability was created in the Multics system. The problem was that the signaller code had been written with other design rules in mind. This is why it is important to: (1) have clear design rules and (2) have approaches and (automated) tools to verify that the implementation meets the design rules. Unfortunately, most operating systems are implemented without clear design rules for security, and few approaches are available to verify compliance with such rules.

Third, Karger and Schell demonstrated that software vulnerabilities, such as the one above, then enable compromise of all Multics security through further "procedural" vulnerabilities. They demonstrate how an attacker can: (1) take control of the Multics patch utility enabling modification of trusted programs; (2) forge the user identification of processes under the control of the attacker; (3) modify the password file; and (4) hide the existence of the attacker by modifying the audit trail and installing backdoors into the system. This work demonstrates many of the challenges that modern operating system designers face of hidden threats, such as *rootkits*. Even if the design is secure and comprehensive, implementation mistakes or poor maintenance decisions can introduce significant vulnerabilities.

3.5   SUMMARY

The Multics designers were the first to tackle the challenge of building an operating system that enables comprehensive enforcement of practical secrecy and integrity requirements. This challenge was just one of several that the designers were faced with, as Multics was also one of the first, structured, time-sharing operating systems as well. As the security analysis shows, the Multics design addressed many facets of building a secure operating system, including defining a reference monitor to enforce a mandatory secrecy policy and developing a protection ring model to protect the integrity of the trusted code, among several innovations. Multics set the foundations for building secure operating systems, but our security analysis and the vulnerability analysis of Karger and Schell show that many difficult issues remain to be addressed. Subsequent work, described in Chapter 6, aimed to address many of these problems, particularly reduction in TCB complexity. First, to clarify the idea of a secure operating system further, we will examine why ordinary operating systems, such as Windows and UNIX, are fundamentally not secure operating systems in Chapter 4.

CHAPTER 4

Security in Ordinary Operating Systems

In considering the requirements of a secure operating system, it is worth considering how far ordinary operating systems are from achieving these requirements. In this chapter, we examine the UNIX and Windows operating systems and show why they are fundamentally not secure operating systems. We first examine the history these systems, briefly describe their protection systems, then we show, using the requirements of a secure operating system defined in Chapter 2, why ordinary operating systems are inherently insecure. Finally, we examine common vulnerabilities in these systems to show the need for secure operating systems and the types of threats that they will have to overcome.

4.1 SYSTEM HISTORIES

4.1.1 UNIX HISTORY

UNIX is a multiuser operating system developed by Dennis Ritchie and Ken Thompson at AT&T Bell Labs [266]. UNIX started as a small project to build an operating system to play a game on an available PDP-7 computer. However, UNIX grew over the next 10 to 15 years into a system with considerable mindshare, such that a variety of commercial UNIX efforts were launched. The lack of coherence in these efforts may have limited the market penetration of UNIX, but many vendors, even Microsoft, had their own versions. UNIX remains a significant operating system today, embodied in many systems, such as Linux, Sun Solaris, IBM AIX, the various BSD systems, etc.

Recall from Chapter 3 that Bell Labs was a member of the Multics consortium. However, Bell Labs dropped out of the Multics project in 1969, primarily due to delays in the project. Ken Thompson adapted some of the ideas of Multics when he initiated the construction of a system that was named as a pun on the Multics system, UNICS (UNIplexed Information and Computing Service). Eventually and mysteriously, the system was renamed UNIX, but the project had begun.

UNIX gained mindshare for a number of reasons. Ritchie rewrote UNIX in his new programming language C which enabled UNIX to be the first portable operating system. This enabled the development of a UNIX community, since lots of people could run UNIX on a variety of different hardware. Next, an application program interface was designed for UNIX which enabled programmers to write application easily, without resorting to assembly language, and these applications ran across the variety of UNIX-supported platforms. Finally, UNIX was truly simplified when compared to Multics. While UNIX adopted many Multics principles, such as hierarchical file systems, virtual memory, and encrypted passwords, UNIX was far simpler. UNIX aimed for a small base program called the *kernel* with a standard interface to simplify the development of applications. As a result, the code size of UNIX (at the time) was smaller than Multics, UNIX performed better, and UNIX was easier to program and administer.

As a streamlined descendant of Multics, UNIX adopted several of the Multics security features, such as password storage, protection ring usage, access control lists, etc., but most were streamlined as well. Since UNIX was not a government-funded project like Multics, it was built with different security goals in mind. For UNIX, the goal was to develop a common platform (e.g., devices and file system) that could be shared by several users. As a result, the security problem became one of *protection* [1], where the goal is to protect the users' data from inadvertent errors in their programs. However, protection does not ensure that secrecy and integrity goals (i.e., security) can be achieved (see Chapter 5). Security enforcement requires that a system's security mechanisms can enforce system security goals even when all the software outside the trusted computing base is malicious. Thus, when UNIX systems were connected to untrusted users via the Internet, a variety of design decisions made for protection no longer applied. As we will discuss, the ordinary UNIX security mechanisms are not capable of enforcing the requirements of a secure operating system. A variety of efforts have aimed to extend or replace the insecure mechanisms for ordinary UNIX systems with mechanisms that may achieve the requirements of a secure operating system (see Chapter 2), as we describe in Chapters 7 and 9.

4.1.2   WINDOWS HISTORY

The history of the Microsoft Windows operating system goes back to the introduction of MS-DOS, which was the original operating system for IBM personal computers introduced in 1981 [24]. MS-DOS was constructed from the Quick and Dirty Operating System (QDOS) built by Tim Paterson that Microsoft purchased from his employer Seattle Computer Products. QDOS was itself based on an early microcomputer operating system called the Control Program for Microcomputers (CP/M) [68, 75]. Compared to other operating systems of the time, such as Multics and UNIX, MS-DOS was a very limited system. It was not a true multitasking system, and did not use many of the features of the x86 processor. Over the next 20 years, Microsoft made improvements to MS-DOS to support more efficient and flexible use of the x86 hardware.

Windows was originally a GUI for MS-DOS, but its visibility soon led to using its name for the subsequent operating systems that Microsoft released. Early Windows systems were based on various versions of MS-DOS, but MS-DOS became less fundamental to the later "Windows 9x" systems. A second, independent line of systems based on the NT kernel emerged starting with the Windows NT 4.0. In 2000, the Windows systems derived from the original MS-DOS codebase were discontinued. At this point, the Windows brand of operating systems dominated the desktop computing market and spanned most computing devices, but the lack of focus on security in Windows operating systems was becoming a significant limitation in these systems.

As the initial focus of the Windows operating system was on microcomputer platforms envisioned for a single user and disconnected from any network, security was not a feature of such

---

[1] Named after the protection system in Lampson's famous paper [176] which achieves the same security goal.

systems. Users administered their systems, uploading new programs as they were purchased. However, the emergence of the world-wide web made connecting Windows computers to the network fundamental to its use, and the networked services that users leveraged, such as email, web clients, easy program download, etc., introduced vulnerabilities that the Windows systems were not designed to counter. The usability model of Windows as a open, flexible, user-administered platform, plus its ubiquity, made it an easy target for attackers. Further, Microsoft was slow to address such threats. In 2000, features were nearly always enabled by default, leading to world-wide compromises due to Windows vulnerabilities (e.g., Code Red and variants [88, 334]). Microsoft has focused with some success on reducing its vulnerabilities through better code development practices [139], code analysis tools [210], and more secure configuration settings. However, improvements in the security features of the Windows operating systems have been less effective. The Windows 2000-based access control system is complex and largely unused [303], the Windows operating system trusted computing base is extremely large (50 million lines of source code in the operating system alone), and recent security enhancements for Windows Vista [152] are both insufficient to provide complete integrity protection [221, 220] and so invasive as to be unpopular [243].

4.2   UNIX SECURITY

We provide a brief outline of a UNIX system prior to examining the security details. Those interested in a comprehensive description of UNIX system concepts are encouraged to read one of the many books on the subject [119, 201, 192].

A running UNIX system consists of an *operating system kernel* and many *processes* each running a program. A protection ring boundary isolates the UNIX kernel from the processes. Each process has its own *address space*, that defines the memory addresses that it can access. Modern UNIX systems define address spaces primarily in terms of the set of *memory pages* that they can access [2]. UNIX uses the concept of a *file* for all persistent system objects, such as secondary storage, I/O devices, network, and interprocess communication. A UNIX process is associated with an *identity*, based on the user associated with the process, and access to files is limited by the process's identity.

UNIX security aims to protect users from each other and the system's trusted computing base (TCB) from all users. Informally, the UNIX TCB consists of the kernel and several processes that run with the identity of the privileged user, root or superuser. These root processes provide a variety of services, including system boot, user authentication, administration, network services, etc. Both the kernel and root processes have full system access. All other processes have limited access based on their associated user's identity.

4.2.1   UNIX PROTECTION SYSTEM
UNIX implements a classical protection system (see Definition 2.1 in Chapter 2), not the secure protection system (see Definition 2.4). As stated in Definition 2.1, a UNIX protection system

---

[2]Segmentation is still supported in most modern processors, but it is not used as the primary access boundary in UNIX systems anymore, as it was in Multics.

consists of a protection state and a set of operations that enable processes to modify that state. Thus, UNIX is a *discretionary access control* (DAC) system. However, UNIX does have some aspects of the secure protection system in Definition 2.4. First, the UNIX protection system defines a *transition state* that describes how processes change between protection domains. Second, the *labeling state* is largely ad hoc. Trusted services associate processes with user identities, but users can control the assignment of permissions to system resources (i.e., files). In the final analysis, these mechanisms and the discretionary protection system are insufficient to build a system that satisfies the secure operating system requirements (see Definition 2.6 in Chapter 2).

Recall that a protection state describes the operations that the system's subjects can perform on that system's objects. The UNIX protection state associates process identities (subjects) with their access to files (objects). Each UNIX process identity consists of a *user id* (UID), a *group id* (GID), and a set of *supplementary groups*. These are used in combination to determine access as described below [3].

All UNIX resources are represented as files. The protection state specifies that subjects may perform read, write, and execute operations on files, with the standard meaning of these operations. While directories are not files, they are represented as files in the UNIX protection state, although the operations have different semantics (e.g., `execute` means search for a directory).

Files are also associated with an owner UID and an owner GID which conveys special privileges to processes with these identities. A process with the owner UID can modify any aspect of the protection state for this file. Processes with either the owner UID and group GID may obtain additional rights to access the file as described below.

The limited set of objects and operations enabled UNIX designers to use a compressed access control list format called *UNIX mode bits*, to specify the access rights of identities to files. Mode bits define the rights of three types of subjects: (1) the file owner UID; (2) the file group GID; and (3) all other subjects. Using mode bits authorization is performed as follows. First, the UNIX authorization mechanism checks whether the process identity's UID corresponds to the owner UID of the file, and if so, uses the mode bits for the owner to authorize access. If the process identity's GID or supplementary groups correspond to the file's group GID, then the mode bits for the group permissions are used. Otherwise, the permissions assigned to all others are used.

Example 4.1.   UNIX mode bits are of the form {owner bits, group bits, others bits} where each element in the tuple consists of a read bit, a write bit, and an execute bit. The mode bits:

`rwxr--r--`

mean that a process with the same UID as the owner can read, write, or execute the file, a process with a GID or supplementary group that corresponds to the file's group can read the file, and others can also only read the file.

---

[3] A process's user identity is actually represented by a set of UIDs for effective, real, and file system access. These details are important to preventing vulnerabilities, see Section 4.2.4, but for clarity we defer their definition until that section.

Suppose a set of files have the following owners, groups, and others mode bits as described below:

```
Name   Owner     Group      Mode Bits
foo    alice     faculty    rwxr--r--
bar    bob       students   rw-rw-r--
baz    charlie   faculty    rwxrwxrwx
```

Then, processes running as alice with the group faculty can read, write, or execute foo and baz, but only read bar. For bar, Alice does not match the UID (bob), nor have the associated group (students). The process has the appropriate owner to gain all privileges for foo and the appropriate group to gain privileges to baz.

As described above, the UNIX protection system is a discretionary access control system. Specifically, this means that a file's mode bits, owner UID, or group GID may be changed by any UNIX processes run by the file's owner (i.e., that have the same UID as the file owner). If we trust all user processes to act in the best interests of the user, then the user's security goals can be enforced. However, this is no longer a reasonable assumption. Nowadays, users run a variety of processes, some of which may be supplied by attackers and others may be vulnerable to compromise from attackers, so the user will have no guarantee that these processes will behave consistently with the user's security goals. As a result, a secure operating system cannot use discretionary access control to enforce user security goals.

Since discretionary access control permits users to change their files owner UID and group GID in addition to the mode bits, file labeling is also discretionary. A secure protection system requires a mandatory *labeling state*, so this is another reason that UNIX systems cannot satisfy the requirements of a secure operating system.

UNIX processes are labeled by trusted services from a set of labels (i.e., user UIDs and group GIDs) defined by trusted administrators, and child processes inherit their process identity from their parent. This mandatory approach to labeling processes with identities would satisfy the secure protection system requirements, although it is rather inflexible.

Finally, UNIX mode bits also include a specification for protection domain transitions, called the setuid bit. When this bit is set on a file, any process that executes the file with automatically perform a protection domain transition to the file's owner UID and group GID. For example, if a root process sets the setuid bit on a file that it owns, then any process that executes that file will run under the root UID. Since the setuid bit is a mode bit, it can be set by the file's owner, so it is also managed in a discretionary manner. A secure protection state requires a mandatory *transition state* describe all protection domain transitions, so the use of discretionary setuid bits is insufficient.

4.2.2   UNIX AUTHORIZATION

The UNIX authorization mechanism controls each process's access to files and implements protection domain transitions that enable a process to change its identity. The authorization mechanism runs in the kernel, but it depends on system and user processes for determining its authorization queries and its protection state. For these and other reasons described in the UNIX security analysis, the UNIX authorization mechanism does not implement a reference monitor. We prove this in the Section 4.2.3 below.

UNIX authorization occurs when files are opened, and the operations allowed on the file are verified on each file access. The requesting process provides the name of the file and the operations that will be requested upon the file in the open system call. If authorized, UNIX creates a *file descriptor* that represents the process's authorized access to perform future operations on the file. File descriptors are stored in the kernel, and only an index is returned to the process. Thus, file descriptors are a form of *capability* (see Chapter 2 for the definition and Chapter 10 for a discussion on capability-based systems). User processes present their file descriptor index to the kernel when they request operations on the files that they have opened.

UNIX authorization controls traditional file operations by mediating file open for read, write, and execute permissions. However, the use of these permissions does not always have the expected effect: (1) these permissions and their semantics do not always enable adequate control and (2) some objects are not represented as files, so they are unmediated. If a user has read access to a file, this is sufficient to perform a wide-variety of operations on the file besides reading. For example, simply via possession of a file descriptor, a user process can perform any ad hoc command on the file using the system calls ioctl or fcntl, as well as read and modify file metadata. Further, UNIX does not mediate all security-sensitive objects, such as network communications. Host firewalls provide some control of network communication, but they do not restrict network communication by process identity.

The UNIX authorization mechanism depends on user-level authentication services, such as login and sshd, to determine the process identity (i.e., UID, GID, and supplementary groups, see Section 4.2.1). When a user logs in to a system, her processes are assigned her login identity. All subsequent processes created in this login session inherit this identity unless there is a domain transition (see below). Such user-level services also need root privileges in order to change the identity of a process, so they run with this special UID. However, several UNIX services need to run as root in order to have the privileges necessary to perform their tasks. These privileges include the ability to change process identity, access system files and directories, change file permissions, etc. Some of these services are critical to the correct operation of UNIX authorization, such as sshd and passwd, but others are not, such as inetd and ftp. However, a UNIX system's trusted computing base must include all root processes, thus risking compromise of security critical services and the kernel itself.

UNIX protection domain transitions are performed by the setuid mechanism. setuid is used in two ways: (1) a root process can invoke the setuid system call to change the UID of a process [4] and (2) a file can have its setuid mode bit set, such that whenever it is executed its identity is set to the owner of the file, as described in Section 4.2.1. In the first case, a privileged process,

---

[4]There are similar commands, such as setgid and setgroups, to change the GID and supplementary groups, respectively.

such as login or sshd, can change the identity of a process. For example, when a user logs in, the login program must change the process identity of the user's first process, her shell, to the user to ensure correct access control. In the second case, the use of the setuid bit on a file is typically used to permit a lower privileged entity to execute a higher privileged program, almost always as root. For example, when a user wishes to change her password, she uses the passwd program. Since the passwd program modifies the password file, it must be privileged, so a process running with the user's identity could not change the password file. The setuid bit on the root-owned, passwd executable's file is set, so when any user executes passwd, the resultant process identity transitions to root. While the identity transition does not impact the user's other processes, the writers of the passwd program must be careful not to allow the program to be tricked into allowing the user to control how passwd uses its additional privileges.

UNIX also has a couple of mechanisms that enable a user to run a process with a reduced set of permissions. Unfortunately, these mechanisms are difficult to use correctly, are only available to root processes, and can only implement modest restrictions. First, UNIX systems have a special principal nobody that owns no files and belongs to no groups. Therefore, a process's permissions can be restricted by running as nobody since it never has owner or group privileges. Unfortunately, nobody, like all subjects, has others privileges. Also, since only root can do a setuid only a superuser process can change the process identity to nobody. Second, UNIX chroot can be used to limit a process to a subtree of the file system [262]. Thus, the process is limited to only its rights to files within that subtree. Unfortunately, a chroot environment must be setup carefully to prevent the process from escaping the limited domain. For example, if an attacker can create /etc/passwd and /etc/shadow files in the subtree, she can add an entry for root, login as this root, and escape the chroot environment (e.g., using root access to kernel memory). Also, a chroot environment can only be setup by a root process, so it is not usable to regular system users. In practice, neither of these approaches has proven to be an effective way to limit process permissions.

4.2.3   UNIX SECURITY ANALYSIS

If UNIX can be a secure operating system, it must satisfy the secure operating system requirements of Chapter 2. However, UNIX fails to meet any of these requirements.

1. **Complete Mediation**: How does the reference monitor interface ensure that all security-sensitive operations are mediated correctly?

   The UNIX reference monitor interface consists of hooks to check access for file or inode permission on some system calls. The UNIX reference monitor interface authorizes access to the objects that the kernel will use in its operations.

   A problem is that the limited set of UNIX operations (read, write, and execute) is not expressive enough to control access to information. As we discussed in Section 4.2.2, UNIX permits modifications to files without the need for write permission (e.g., fcntl).

2. **Complete Mediation**: Does the reference monitor interface mediate security-sensitive operations on all system resources?

   UNIX authorization does not provide complete mediation of all system resources. For some objects, such as network communications, UNIX itself provides no authorization at all.

3. **Complete Mediation**: How do we verify that the reference monitor interface provides complete mediation?

   Since the UNIX reference monitor interface is placed where the security-sensitive operations are performed, it difficult to know whether all operations have been identified and all paths have been mediated. No specific approach has been used to verify complete mediation.

4. **Tamperproof**: How does the system protect the reference monitor, including its protection system, from modification?

   The reference monitor and protection system are stored in the kernel, but this does not guarantee tamper-protection. First, the protection system is discretionary, so it may be tampered by any running process. Untrusted user processes can modify permissions to their user's data arbitrarily, so enforcing security goals on user data is not possible.

   Second, the UNIX kernel is not as protected from untrusted user processes as the Multics kernel is. Both use protection rings for isolation, but the Multics system also explicitly specifies *gates* for verifying the legality of the ring transition arguments. While UNIX kernels often provide procedures to verify system call arguments, such procedures are may be misplaced.

   Finally, user-level processes have a variety of interfaces to access and modify the kernel itself above and beyond system calls, ranging from the ability to install kernel modules to special file systems (e.g., /proc or *sysfs*) to interfaces through netlink sockets to direct access to kernel memory (e.g., via the device file/dev/kmem). Ensuring that these interfaces can only be accessed by trusted code has become impractical.

5. **Tamperproof**: Does the system's protection system protect the trusted computing base programs?

   In addition to the kernel, the UNIX TCB consists of *all* root processes, including all processes run by a user logged in as a root user. Since these processes could run any program, guaranteeing the tamper-protection of the TCB is not possible. Even ignoring root users, the amount of TCB code is far too large and faces far too many threats to claim a tamperproof trusting computing base. For example, several root processes have open network ports that may be used as avenues to compromise these processes. If any of these processes is compromised, the UNIX system is effectively compromised as there is no effective protection among root processes.

Also, any `root` process can modify any aspect of the protection system. As we show below, UNIX `root` processes may not be sufficiently trusted or protected, so unauthorized modification of the protection system, in general, is possible. As a result, we cannot depend on a tamperproof protection system in a UNIX system.

6. **Verifiable**: What is basis for the correctness of the system's TCB?

   Any basis for correctness in a UNIX system is informal. The effectively unbounded size of the TCB prevents any effective formal verification. Further, the size and extensible nature of the kernel (e.g., via new device drivers and other kernel modules) makes it impractical to verify its correctness.

7. **Verifiable**: Does the protection system enforce the system's security goals?

   Verifiability enforcement of security goals is not possible because of the lack of complete mediation and the lack of tamperproofing. Since we cannot express a policy rich enough to prevent unauthorized data leakage or modification, we cannot enforce secrecy or integrity security goals. Since we cannot prove that the TCB is protected from attackers, we cannot prove that the system will be remain able to enforce our intended security goals, even if they could be expressed properly.

4.2.4 UNIX VULNERABILITIES

A secure operating system must protect its trusted computing base from compromise in order to implement the reference monitor guarantees as well. In this section, we list some of the vulnerabilities that have been found in UNIX systems over the years that have resulted in the compromise of the UNIX trusted computing base. This list is by no means comprehensive. Rather, we aim to provide some examples of the types of problems encountered when the system design does not focus on protecting the integrity of the trusted computing base.

Network-facing Daemons   UNIX has several `root` (i.e., TCB) processes that maintain network ports that are open to all remote parties (e.g., `sshd`, `ftpd`, `sendmail`, etc.), called *network-facing daemons*. In order to maintain the integrity of the system's trusted computing base, and hence achieve the reference monitor guarantees, such process must protect themselves from such input. However, several vulnerabilities have been reported for such processes, particularly due to buffer overflows [232, 318], enabling remote attackers to compromise the system TCB. Some of these daemons have been redesigned to remove many of such vulnerabilities (e.g., Postfix [317, 73] as a replacement for `sendmail` and privilege-separated SSH [251]), but a comprehensive justification of integrity protection for the resulting daemons is not provided. Thus, integrity protection of network-facing dameons in UNIX is incomplete and ad hoc.

Further, some network-facing daemons, such as remote login daemons (e.g., `telnet`, `rlogin`, etc.) `ftpd`, and NFS, puts an undo amount of trust in the network. The remote login daemons and ftpd are notorious for sending passwords in the clear. Fortunately, such daemons have been obsoleted or replaced by more secure versions (e.g., vsftpd for ftpd). Also, NFS is notorious for accepting any response to a remote file system request as being from a legitimate server [38]. Network-facing daemons must additionally protect the integrity of their secrets and authenticate the sources of remote data whose integrity is crucial to the process.

Rootkits   Modern UNIX systems support extension via kernel modules that may be loaded dynamically into the kernel. However, a malicious or buggy module may enable an attacker to execute code in the kernel, with full system privileges. A variety of malware packages, called *rootkits*, have been created for taking advantage of kernel module loading or other interfaces to the kernel available to root processes. Such rootkits enable the implementation of attacker function and provide measures to evade from detection. Despite efforts to detect malware in the kernel [244, 245], such rootkits are difficult to detect, in general, [17].

Environment Variables   UNIX systems support *environment variables*, system variables that are available to processes to convey state across applications. One such variable is LIBPATH whose value determines the search order for dynamic libraries. A common vulnerability is that an attacker can change LIBPATH to load an attacker-provided file as a dynamic library. Since environment variables are inherited when a child process is created, an untrusted process can invoke a TCB program (e.g., a program file which setuid's to root on invocation, see Section 4.2.2) under an untrusted environment. If the TCB process depends on dynamic libraries and does not set the LIBPATH itself, it may be vulnerable to running malicious code. As many TCB programs can be invoked via setuid, this is a widespread issue.

Further, TCB programs may be vulnerable to any input value supplied by an untrusted process, such as malicious input arguments. For example, a variety of program permit the caller to define the configuration file of the process. A configuration file typically describes all the other places that the program should look for inputs to describe how it should function, sometimes including the location of libraries that it should use and the location of hosts that provide network information. If the attack can control the choice of a program's configuration file, she often has a variety of ways to compromise the running process. Any TCB program must ensure their integrity regardless of how they are invoked.

Shared Resources   If TCB processes share resources with untrusted processes, then they may be vulnerable to attack. A common problem is the sharing of the /tmp directory. Since any process can create files in this directory, an untrusted process is able to create files in this directory and grant other processes, in particular a TCB process, access to such files as well. If the untrusted process can guess the name of TCB process's /tmp file, it can create this file in advance, grant access to the TCB process, and then have access itself to a TCB file. TCB processes can prevent this problem by checking for the existence of such files upon creation (e.g., using the O_CREAT flag). However, programmers have been prone to forget such safeguards. TCB process must take care when using any objects shared by untrusted processes.

Time-of-Check-to-Time-of-Use (TOCTTOU)   Finally, UNIX has been prone to a variety of attacks where untrusted processes may change the state of the system between the time an operation is authorized and the time that the operation is performed. If such a change enables an untrusted process to access a file that would not have been authorized for, then this presents a vulnerability. The attack was first identified by Dilger and Bishop [30] who gave it the moniker *time-of-check-to-time-of-use attacks* or TOCTTOU attacks. In the classical example, a root process uses the system call access to determine if the user for whom the process is running (e.g., the process was initiated by a setuid) has access to a particular file /tmp/X. However, after the access system call authorizes the file access and before the file open, the user may change the binding between the file name and the actual file object (i.e., inode) accessed. This can be done by change the file /tmp/X to a symbolic link to the target file /etc/shadow. As a result, UNIX added a flag, so the open request could prevent traversal via symbolic links. However, the UNIX file system remains susceptible to TOCTTOU attacks because the mapping between file names and actual file objects (inodes) can be manipulated by the untrusted processes.

As a result of the discretionary protection system, the size of the system TCB, and these types of vulnerabilities, converting a UNIX system to a secure operating system is a significant challenge. Ensuring that TCB processes protect themselves, and thus protect a reference monitor from tampering, is a complex undertaking as untrusted processes can control how TCB processes are invoked and provide inputs in multiple ways: network, environment, and arguments. Further, untrusted processes may use system interfaces to manipulate any shared resources and may even change the binding between object name and the actual object. We will discuss the types of changes necessary to convert an ordinary UNIX system to a system that aims to satisfy the secure operating system definition in Chapters 7 and 9, so we will see that several fundamental changes are necessary to overcome these problems. Even then, the complexity of UNIX systems and their trusted computing base makes satisfying the tamperproof and verifiability requirements of the reference monitor concept very difficult.

4.3   WINDOWS SECURITY

In this section, we will show that Windows operating systems also fail to meet the requirements of a secure operating system. This section will be much briefer than the previous examination of UNIX as many of the concepts are similar. For example, Windows also supports processes with their own address spaces that are managed by a ring-protected kernel. For a detailed description of the Windows access control system examined in this section, circa Windows 2000, see Swift et al. [303].

4.3.1   WINDOWS PROTECTION SYSTEM

The Windows 2000 protection system [5], like the UNIX protection system, provides a discretionary access control model for managing protection state, object labeling, and protection domain transitions. The two protection systems manly differ in terms of flexibility (e.g., the Windows system is extensible) and expressive power (e.g., the Windows system enables the description of a wider variety of policies). Unfortunately, when we compare the Windows protection system to the definition of a secure protection system, we find that improvements in flexibility and expressive power actually make the system more difficult to secure.

Specifically, the Windows protection system differs from UNIX mainly in the variety of its objects and operations and the additional flexibility it provides for assigning them to subjects. When the Windows 2000 access control model was being developed, there were a variety of security systems being developed that provided administrators with extensible policy languages that permitted flexible policy specification, such as the Java 2 model [117]. While these models address some of the shortcomings of the UNIX model by enabling the expression of any protection state, they do not ensure a secure system.

Subjects in Windows are similar to subjects in UNIX. In Windows, each process is assigned a *token* that describes the process's identity. A process identity consists of user security identifier (principal SID, analogous to a UNIX UID), a set of group SIDs (rather than a single UNIX GID and a set of supplementary groups), a set of alias SIDs (to enable actions on behalf of another identity), and a set of privileges (ad hoc privileges just associated with this token). A Windows identity is still associated with a single user identity, but a process token for that user may contain any combination of rights.

Unlike UNIX, Windows objects can belong to a number of different data types besides files. In fact, applications may define new data types, and add them to the *active directory*, the hierarchical name space for all objects known to the system. From an access control perspective, object types are defined by their set of operations. The Windows model also supports a more general view of the operations that an object type may possess. Windows defines up to 30 operations per object type, including some operations that are specific to the data type [74]. This contrasts markedly with the read, write, and execute operations in the UNIX protection state. Even for file objects, the Windows protection system defines many more operations, such as operations to access file attributes and synchronize file operations. In addition, application may add new object types and define their own operations.

The other major difference between a Windows and UNIX protection state is that Windows supports arbitrary access control lists (ACLs) rather than the limited mode bits approach of UNIX. A Windows ACL stores a set of access control entries (ACEs) that describe which operations an SID (user, group, or alias) can perform on that object [6]. The operations in an ACE are interpreted based on the object type of the target object. In Windows, ACEs may either grant or deny an operation.

---

[5]We simply refer to this as the Windows protection system for the rest of the chapter.
[6]Remember that access control lists are stored with the object, and state which subjects can access that object.

Thus, Windows uses negative access rights, whereas UNIX does not, generating some differences in their authorization mechanisms.

Example 4.2.   Figure 4.1 shows an example ACL for an object foo. foo's ACL contains three ACEs. The field *principal SID* specifies the SID to which the ACE applies. These ACE apply to

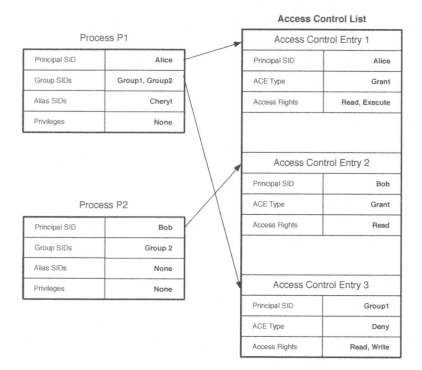

Figure 4.1: Windows Access Control Lists (ACLs) and process tokens for Examples 4.2 and 4.3

the SIDs Alice, Bob, and Group1. The other two important fields in an ACE are its *type* (grant or deny) and the *access rights* (a bitmask). The Alice and Bob ACEs grant rights, and the Group1 ACE denies access to certain rights. The access rights bitmask is interpreted based on the *object type* field in the ACE. We describe how the ACL is used in authorization in the next section.

4.3.2   WINDOWS AUTHORIZATION

Windows authorization queries are processed by a specific component called the *Security Reference Monitor* (SRM). The SRM is a kernel component that takes a process token, an object SID, and a set of operations, and it returns a boolean result of an authorization query. The SRM uses the object SID to retrieve its ACL from which it determines the query result.

Because of the negative permissions, the way that the SRM processes authorization queries is more complicated than in the UNIX case. The main difference is that the ACEs in an ACL are ordered, and the ACEs are examined in that order. The SRM searches the ACEs until it finds a set of ACEs that permits the operation or a single ACE that denies the operation. If an ACE grants the necessary operations [7], then the request is authorized. However, if a deny ACE is encountered that includes one of the requested operations, then the entire request is denied.

Example 4.3. Returning to Example 4.2 above, the ACEs of the object's ACL are ordered as shown in Figure 4.1. Note that the ACE field for access rights is really a bitmap, but we list the operations to simplify understanding. Further, we specify the process tokens for two processes, P1 and P2. Below, we show the authorization results for a set of queries by these processes for the target object.

```
P1, read: ok
P1, read, write: no
P2: read: ok
P2: read, write: no
```

Both P1 and P2 can read the target object, but neither can write the object. P1 cannot write the object because the P1 token include `Group1` which matches the deny ACE for writing. P2 cannot write the object because the ACE for Bob does not permit writing.

Mediation in Windows is determined by a set of object managers. Rather than a monolithic set of system calls to access homogeneous objects (i.e., files) in UNIX, each object type in Windows has an object manager that implements the functions of that type. While the Windows object managers all run in the kernel, the object managers are independent entities. This can be advantageous from a modularity perspective, but the fact that object managers may extend the system presents some challenges for mediation. We need to know that each new object manager mediates all operations and determines the rights for those operations correctly. There is no process for ensuring this in Windows.

In Windows, the trusted computing base consists of all system services and processing running as a trusted user identity, such as `Administrator` [8]. Windows provides a `setuid`-like mechanism for invoking Windows *Services* that run at a predefined privilege, at least sufficient to support all clients. Thus, vulnerabilities in such services would lead to system compromise. Further, the ease of software installation and complexity of the discretionary Windows access control model often result in users running as `Administrator`. In this case, any user program would be able to take control of the system. This is often a problem on Windows systems. With the release of Windows Vista, the Windows model is extended to prevent programs downloaded from the Internet from

---

[7]It may take multiple ACEs to grant all the requested operations, so this refers to the ACE that grants whatever remaining operations were requested.

[8]In addition, these services and processes may further depend on non-Administrator processes, which would make the system TCB even less secure.

automatically being able to write Windows applications and the Windows system, regardless of the user's process identity [152]. While this does provide some integrity protection, it does not fully protect the system's integrity. It prevents low integrity processes from writing to high integrity files, but does not prevent invocation, malicious requests, or spoofing the high integrity code into using a low integrity file. See Chapter 5 for the integrity requirements of a secure operating system.

Windows also provides a means for restricting the permissions available to a process flexibly, called *restricted contexts*. By defining a restricted context for a process, the permissions necessary to perform an operation must be available to both the process using its token and to the restricted context. That is, the permissions of a process running in a restricted context are the *intersection* of the restricted context and the process's normal permissions. Since a restricted context may be assigned an arbitrary set of permissions, this mechanism is much more flexible than the UNIX option of running as nobody. Also, since restricted contexts are built into the access control system, it less error-prone than and `chroot`. Nonetheless, restricted contexts are difficult for administrators to define correctly, so they are not used commonly, and not at all by the user community.

4.3.3 WINDOWS SECURITY ANALYSIS

Despite the additional expressive power offered by the Windows access control model, it also does not satisfy any of the reference monitor guarantees either. Although Windows can express any combination of permissions, it becomes more difficult to administer. In my informal polls, no users use the Windows permission model at all, whereas most at least were aware of how to use the UNIX model (although not always correctly). Windows is effectively no more or less secure than ordinary UNIX—they are both insecure.

1. **Complete Mediation**: How does the reference monitor interface ensure that all security-sensitive operations are mediated correctly?

   In Windows, mediation is provided by object managers. Without the source code, it is difficult to know where mediation is performed, but we would presume that object managers would authorize the actual objects used in the security-sensitive operations, similarly to UNIX.

2. **Complete Mediation**: Does the reference monitor interface mediate security-sensitive operations on all system resources?

   Object managers provide an opportunity for complete mediation, but provide no guarantee of mediation. Further, the set of managers may be extended, resulting in the addition of potentially insecure object managers. Without a formal approach that defines what each manager does and how it is to be secured, it will not be possible to provide a guarantee of complete mediation.

3. **Complete Mediation**: How do we verify that the reference monitor interface provides complete mediation?

   As for UNIX, no specific approach has been used to verify complete mediation.

4. **Tamperproof**: How does the system protect the reference monitor, including its protection system, for modification?

Windows suffers from the same problems as UNIX when it comes to tampering. First, the protection system is discretionary, so it may be tampered by any running process. Untrusted user processes can modify permissions to their user's data arbitrarily, so enforcing security goals on user data is not possible. Since users have often run as `Administrator` to enable ease of system administration, any aspect of the protection system may be modified.

Second, there are limited protections for the kernel itself. Like UNIX, a Windows kernel can be modified through kernel modules. In Microsoft Vista, a code signing process can be used to determine the certifier of a kernel module (i.e., the signer, not necessarily the writer of the module). Of course, the administrator (typically an end user) must be able to determine the trustworthiness of the signer. Security procedures that depend on the decision-making of users are often prone to failure, as users are often ignorant of the security implications of such decisions. Also, like UNIX, the Windows kernel also does not define protections for system calls (e.g., Multics *gates*).

5. **Tamperproof**: Does the system's protection system protect the trusted computing base programs?

The TCB of Windows system is no better than that of UNIX. Nearly any program may be part of the Windows TCB, and any process running these programs can modify other TCB programs invalidating the TCB.

Like UNIX, any compromised TCB process can modify the protection system invalidating the enforcement of system security goals, and modify the Windows kernel itself through the variety of interfaces provided to TCB processes to access kernel state.

Unlike UNIX, Windows provides APIs to tamper with other processes in ways that UNIX does not. For example, Windows provides the `CreateRemoteThread` function, which enables a process to initiate a thread in another process [207]. Windows also provides functions for writing a processes memory via `OpenProcess` and `WriteProcessMemory`, so one process can also write the desired code into that process prior to initiating a thread in that process. While all of these operations require the necessary access rights to the other process, usually requiring a change in privileges necessary for debugging a process (via the `AdjustTokenPrivileges`). While such privileges are typically only available to processes under the same SID, we must verify that these privileges cannot be misused in order to ensure tamper-protection of our TCB.

6. **Verifiable**: What is basis for the correctness of the system's trusted computing base?

As for UNIX, any basis for correctness is informal. Windows also has an unbounded TCB and extensible kernel system that prevent any effective formal verification.

7. **Verifiable**: Does the protection system enforce the system's security goals?

The general Windows model enables any permission combination to be specified, but no particular security goals are defined in the system. Thus, it is not possible to tell whether a system is secure. Since the model is more complex than the UNIX model and can be extended arbitrarily, this makes verifying security even more difficult.

4.3.4   WINDOWS VULNERABILITIES

Not surprisingly given its common limitations, Windows suffers from the same kinds of vulnerabilities as the UNIX system (see Section 4.2.4). For example, there are books devoted to constructing Windows rootkits [137]. Here we highlight a few vulnerabilities that are specific to Windows systems or are more profound in Windows systems.

The Windows Registry   The Windows Registry is a global, hierarchical database to store data for all programs [206]. When a new application is loaded it may update the registry with application-specific, such as security-sensitive information such as the paths to libraries and executables to be loaded for the application. While each registry entry can be associated with a security context that limits access, such limitations are generally not effectively used. For example, the standard configuration of *AOL* adds a registry entry that specifies the name of a Windows library file (i.e., DLL) to be loaded with AOL software [120]. However, the permissions were set such that any user could write the entry.

This use of the registry is not uncommon, as vendors have to ensure that their software will execute when it is downloaded. Naturally, a user will be upset if she downloads some newly-purchased software, and it does not execute correctly because it could not access its necessary libraries. Since the application vendors cannot know the ad hoc ways that a Windows system is administered, they must turn on permissions to ensure that whatever the user does the software runs. If the registry entry is later used by an attacker to compromise the Windows system, that is not really the application vendor's problem—selling applications is.

Administrator Users   We mentioned in the Windows security evaluation that traditionally users ran under the identity `Administrator` or at least with administrative privileges enabled. The reason for this is similar to the reason that broad access is granted to registry entries: the user also wants to be sure that they can use what function is necessary to enable the system to run. If the user downloads some computer game, the user would need special privileges to install the game, and likely need special privileges to run the device-intensive game program. The last thing the user wants is to have to figure out why the game will not run, so enabling all privileges works around this issue.

UNIX systems are generally used by more experienced computer users who understand the difference between installing software (e.g., run `sudo`) and the normal operation of the computer. As a result, the distinction between `root` users and `sudo` operations has been utilized more effectively in UNIX.

Enabled By Default   Like users and software vendors, Windows deployments also came with full permissions and functionality enabled. This resulted in the famous Code Red worms [88] which attacked the SQL server component of the Microsoft IIS web server. Many people who ran IIS did not have an SQL server running or even knew that the SQL server was enabled by default in their IIS system. But in these halcyon times, IIS web servers ran with all software enabled, so attackers could send malicious requests to SQL servers on any system, triggering a buffer overflow that was the basis for this worm's launch. Subsequent versions of IIS are now "locked down" [9], such that software has to be manually enabled to be accessible.

4.4   SUMMARY

This investigation of the UNIX and Windows protection systems shows that it is not enough just to design an operating system to enforce security policies. Security enforcement must be comprehensive (i.e., mediate completely), mandatory (i.e., tamperproof), and verifiable. Both UNIX and Windows originated in an environment in which security requirements were very limited. For UNIX, the only security requirement was *protection* from other users, and for Windows, users were assumed to be mutually-trusted on early home computers. The connection of these systems to untrusted users and malware on the Internet changed the security requirements for such systems, but the systems did not evolve.

Security enforcement requires that a system's security mechanisms can enforce system security goals even when any of the software outside the trusted computing base may be malicious. This assumption is required in today's world where any network request may be malicious or any user process may be compromised. A system that enforces security goals must implement a mandatory protection system, whereas these system's implement discretionary protection that can be modified and invalidated by untrusted processes. A system that enforces security goals must identify and mediate all security-sensitive operations, whereas these systems have incomplete and informal mediation of access. Finally, a system that enforces security goals must be tamperproof, and these systems have unbounded TCBs that provide many unchecked opportunities for untrusted processes to tamper with the kernel and other TCB software. When we consider secure commercial systems in Chapters 7 and 9, we will see that significant changes are necessary, but it is still difficult to undo fully the legacy of insecurity in these systems.

---

[9]Features that are not required are disabled by default. Bastille Linux performs a similar role to lock down all services in Linux systems [20].

CHAPTER 5

Verifiable Security Goals

In this chapter, we examine access control models that satisfy the *mandatory protection system* of Definition 2.4 in Chapter 2. A mandatory protection system provides a tamperproof description of the system's access control policy. A mandatory protection system consists of: (1) a *protection state* that defines the operations that a fixed set of subject labels can perform on a fixed set of object labels; (2) a *labeling state* that maps system processes and resources to their subject and object labels, respectively; and (3) a *transition state* that defines the legal ways that system processes and resources may be assigned to new labels. As such, it manages the access rights that all system processes will ever obtain.

A mandatory protection system is necessary for a secure operating system to implement two of the requirements of the *reference monitor concept* as defined in Definition 2.6: tamperproofing and verifiability. A mandatory protection system is tamperproof from untrusted processes as the system defines the labeling of subjects and objects, transitions in these labels, and the resulting protection state. Because these access control models in mandatory protection systems only allow the system to modify the protection state, they are called *mandatory access control models* [179] [1]. Also, a mandatory protection system describes the access control policy that we use to verify the enforcement of the system's security goals. Often, such models support the definition of policies that describe concrete security goals, which we will define below.

In this chapter, we describe mandatory protection systems and the security goals that they imply. In general, a secure operating system should ensure the enforcement of secrecy goals, including for covert channels, and integrity goals. We present the basic concept for expressing secrecy and integrity goals, information flows, and then describe models for expressing these goals in mandatory access control policies. The models that we present here mainly focus on either secrecy or integrity, not both, but there are several mandatory models that combine both facets, such as Type Enforcement [33], Chinese Wall [37], and Caernarvon [278]. Unifying secrecy and integrity effectively in systems is an ongoing challenge in practice, as we will see.

5.1 INFORMATION FLOW

Secure operating systems use *information flow* as the basis for specifying secrecy and integrity security requirements. Conceptually, information flow is quite simple.

Definition 5.1. An *information flow* occurs between a subject $s \in S$ and an object $o \in O$ if the subject performs a *read* or *write* operation on the object. The information flow $s \rightarrow o$ is from the

---

[1] Historically, the term *mandatory access control model* has been used to describe specific access control models, such as the multilevel secrecy models, but we use the broader definition in this book that is based on how the policies in these models are administered.

subject to the object if the subject writes to the object. The information flow $s \leftarrow o$ is from the object to the subject if the subject reads from the object.

Information flow represents how data moves among subjects and objects in a system. When a subject (e.g., process) reads from an object (e.g., a file), the data from the object flows into the subject's memory. If there are secrets in the object, then information flow shows that these secrets may flow to the subject when the subject reads the object. However, if the subject holds the secrets, then information flow also can show that the subject may leak these secrets if the subject writes to the object.

Note that every operation on an object is either an information flow read (i.e., extracts data from the object), an information flow write (i.e., updates the object with new data), or a combination of both. For example, *execute* reads the data from a file to prepare it for execution, so the process reads from the file. When we *delete* a file from a directory, the directory's set of files is changed. The result is that an entire protection state of a mandatory protection system can be represented by a set of information flow reads and writes.

Thus, any protection state can be represented by an information flow graph.

Definition 5.2.   An *information flow graph* for a protection state is a directed graph $G = (V, E)$ where: (1) the set of vertices V consists of the union of the set of subjects and set of objects in the protection state and (2) the set of directed edges E consists of each read and write information flow in the protection state.

An information flow graph for a protection state can be constructed as follows. First, we create a vertex for each subject and object in the protection state. Then, we add the information flow edges. To do this, we determine whether each operation in the protection state results in a read, write, or combination information flow. Then, we add an information flow edge from a subject vertex to an object vertex when the subject has permission to a write information flow operation for the object in the protection state. Likewise, we add an information flow edge from an object vertex to a subject vertex when the subject has permission to a read information flow operation in the protection state.

Example 5.3.   Consider the access matrix shown in Figure 5.1. It defines the set of operations that the subjects $S1$, $S2$, and $S3$ can perform on objects $O1$ and $O2$. Note that some operations, such as `append`, `getattr`, and `ioctl` have to be mapped to their resultant information flows, write, read, and both, respectively. As a result, this access matrix represents the corresponding information flow graph shown in Figure 5.1.

Information flow is used in secure operating systems as an approximation for secrecy and integrity. For secrecy, the information flow edges in the graph indicate all the paths by which data may be *leaked*. We can use the graph to determine whether a secret object o may be leaked to an

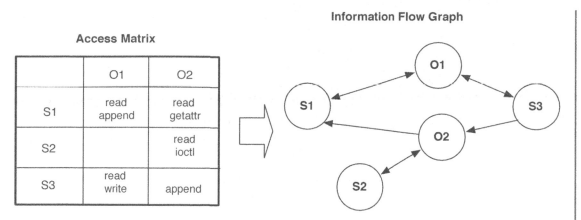

Figure 5.1: The information flow graph on the right represents the information flows described by the access control matrix on the left.

unauthorized subject s. If there exists a path in the information flow graph from o to s, then there is an unauthorized leak in the corresponding protection state.

For integrity, we require that no high integrity subject *depends* on any low integrity subjects or objects. We can use the graph to determine whether a high integrity subject $s1$ receives input from a low integrity subject $s2$ (e.g., an attacker). If there exists a path in the information flow graph from $s2$ to $s1$, then the high integrity subject $s1$ receives input from the low integrity subject $s2$. When a high integrity subject receives input from a low integrity subject, it is assumed that it depends on that low integrity subject.

5.2   INFORMATION FLOW SECRECY MODELS

For information flow secrecy, we want to ensure that no matter which programs a user runs, she cannot leak information to an unauthorized subject. The classical problem is that the user may be coerced into running a program that contains malware that actively wants to leak her information. For example, a *Trojan horse* is a type of malware that masquerades as a legitimate program, but contains a malicious component that tries to leak data to the attacker.

The access control models of UNIX and Windows cannot prevent such an attack, because: (1) they do not account for all the information flows that may be used to leak information and (2) the policies are discretionary, so the malware can modify the policy to introduce illegal information flows. First, UNIX and Windows policies often define some files shared among all users, but any user's secret data may also be leaked through such public files by well-crafted malware. For example, the UNIX model permits the sharing of files with all users by granting read access to the others. However, if a user has write access to any file that others can read, then the malware can leak secrets by writing them to this file. Second, discretionary access control (DAC) protection systems, such as those used by UNIX and Windows, permit the users to modify access control policies. However, any program that the user runs can modify the permission assignments to files that the user owns. Thus, any of the user's files may be leaked by malware simply by changing the file permissions.

The security policy models used in mandatory protection systems aim to solve these problems. First, they explicitly restrict information flows to not leak secret data. Second, such models do not permit users or their programs to modify the information flows permitted. In addition to the secrecy models presented here, see the High-Water Mark model of Weissman [329].

5.2.1   DENNING'S LATTICE MODEL

Denning refined the general information flow graph to express information flow secrecy requirements [70, 271] based partly on the work of Fenton [93].

Definition 5.4.   An *information flow model* is a quintuple $\{N, P, SC, \oplus, \rightarrow\}$, where: (1) N is the set of storage *objects* (e.g., files); P is the set of *subjects* (e.g., processes) that cause information flows; SC is a set of *security classes*; (2) $\rightarrow \subseteq SC \times SC$ is a binary *can-flow* relation on SC; and (3) $\oplus$: $SC \times SC \rightarrow SC$ is a binary *join* operator on SC.

In a information flow model, each subject and object is assigned a security class. Secure classes are labels in the mandatory protection system defined in Definition 2.4, and both subjects and objects may share security classes. For example, a subject and object may both be assigned to security class X. However, another subject, to whom X data must not be leaked, is assigned security class Y. The *can-flow* relation $\rightarrow$ defines the legal information flows in the model. That is, $Y \rightarrow X$ specifies that information at security class Y can flow to subjects and objects in security class X [2]. Since we have a secrecy requirement that information in security class X not be leaked to subjects and objects of security class Y, $X \nrightarrow Y$. The *join* operator determines the security class that results from combining data of two distinct security classes $X \oplus Y = Z$. In this case, the combination of data from X and Y security classes is labeled Z.

Example 5.5.   Figure 5.2 shows two information flow model policies. In (a), this policy isolates users $u_1, u_2, u_3, ..., u_i$ by assigning them to distinct security classes. Any data in security class u_i cannot be read or written by any process running with security class u_j where $i \neq j$.

Figure 5.2 (b) shows an information flow model policy that totally orders security classes, such that data in higher classes will not be leaked to lower security class. These security classes represent the traditional governmental secrecy classes, *top-secret*, *secret*, *confidential*, and *unclassified*. In this policy, *top-secret* data may not be read by processes running in the lower classes in the information flow model policy. Further, processes running in *top-secret* may not write to objects in the lower security classes. Lower security classes, such as *secret*, may have information flow by permitting higher security classes (e.g., *top-secret*) to read their data or by writing up to objects in the higher

---

[2]Infix notation.

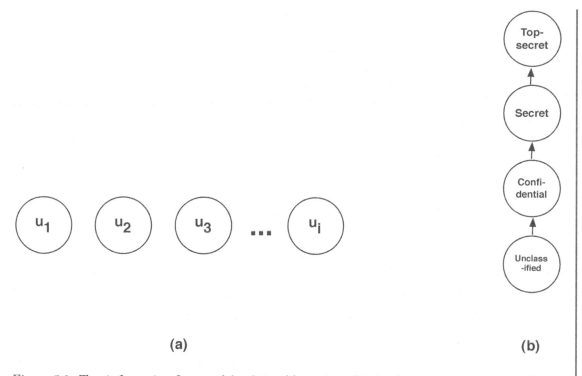

(a)   **(b)**

Figure 5.2: Two information flow model policies: (a) consists of isolated security class where no information flows among them and (b) is a totally-ordered sequence of security classes where information flows upwards only.

security classes. Since processes in the lower security classes do not even know the name of objects in higher security classes, such writing is implemented by a *polyinstantiated* file system where the files have instances at each security level, so the high security process can read the lower data and update the higher secrecy version without leaking whether there is a higher secrecy version of the file.

Such information flow model policies actually can be represented by a finite *lattice*. Denning defines four axioms required for lattice policies.

Definition 5.6. An information flow model forms a *finite lattice* if it satisfies the following axioms.

1. The set of security classes SC is finite.

2. The *can-flow* relation $\to$ is a partial order on SC.

3. SC has a lower bound with respect to $\to$.

4. The *join* operator ⊕ is a totally defined least upper bound operator.

The key result comes from axiom 4. In a finite lattice, the *join* operator is defined for any combination of security classes. Thus, for $X_1 \oplus X_2 \oplus ... \oplus X_n = Z$, the security class Z that results from a combination of data from any security classes in SC must also be in SC. For lattice policies, the results of any join operation can be modeled because the security class of the operation can always be computed.

Note that the Example 5.5(b) satisfies the Denning axioms, so it is a finite lattice. Any combination of data can be assigned a security class. For example, when a *secret* process generates data read from *confidential* and *unclassified* inputs, the data generated is labeled *secret*, which is the least upper bound of the three security classes. However, we can see that Example 5.5(a) does not satisfy axioms 3 and 4, so we cannot label the results from any operation that uses the data from two different users.

Finally, a useful concept is the inverse of the *can-flow* relation, called the *dominance* relation. Dominance is typically used in the security literature.

Definition 5.7. $A \geq B$ (read as A *dominates* B) if and only if $B \rightarrow A$. The *strictly dominates* relation $>$ is defined by $A > B$ if and only if $A \geq B$ and $A \neq B$. We say that A and B are *comparable* if $A \geq B$ or $B \geq A$. Otherwise, A and B are *incomparable*.

Dominance indicates which security class is more sensitive (i.e., contains data that is more secret). From a security perspective, dominance defines the information flows that are not allowed. That is, if $A > B$, then A's data must not flow to B or this constitutes a leak.

5.2.2   BELL-LAPADULA MODEL

The most common information flow model in secure operating systems for enforcing secrecy requirements is the Bell-LaPadula (BLP) model [23]. There are a variety of models associated with Bell and LaPadula, but we describe a common variant here, known as the Multics interpretation. This BLP model is a finite lattice model where the security classes represent two dimensions of secrecy: *sensitivity level* and *need-to-know*. The sensitive level of data is a total order indicating secrecy regardless of the type of data. In the BLP model, these levels consist of the four governmental security classes mentioned previously: *top-secret*, *secret*, *confidential*, and *unclassified*. However, it was found that not everyone with a particular security class "needs to know" all the information labeled for that class. The BLP model includes a set of *categories* that describe the topic areas for data, defining the need-to-know access. The BLP model assigns a sensitivity level that defines the secrecy level that the subject is authorized for, and also a set of categories, called a *compartment*, to each subject and object. The combination of sensitivity level and compartment for a subject are often called its *clearance*. For objects, their combination of sensitivity level and compartment are called its *access class*.

Example 5.8.   Figure 5.3 shows a Bell-LaPadula policy with two sensitivity levels and three categories. The edges show the direction of information flow authorized by the Bell-LaPadula policy. If

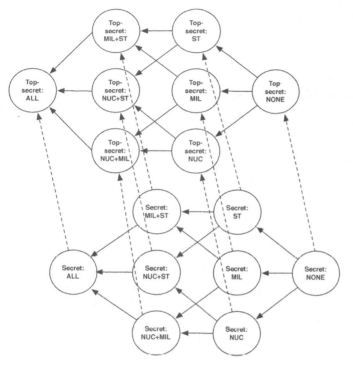

Figure 5.3: This a Haase diagram (with the information flow direction added in edges) of a Bell-LaPadula policy consisting of two sensitivity levels (*top-secret* and *secret* where *top-secret* dominates) and three categories (*NUC*, *MIL*, and *ST*). The edges show the information flows authorized by the Bell-LaPadula model for this lattice.

a subject is cleared for top-secret:MIL, it is able to read from this class, top-secret:none, and secret:MIL. However, information cannot flow to the top-secret:MIL class from top-secret:MIL+ST or others that include categories besides MIL. Even information that is labeled with the secret sensitivity level, but has additional categories may not flow to top-secret:MIL. Of course, subjects at the top-secret:MIL clearance can write to any top-secret class that includes the category MIL, but none of secret classes. The latter is not possible because the sensitivity level top-secret dominates or is incomparable to any secret class. Writes may only be allowed to classes that dominate the subject's clearance.

The BLP model defines two key properties for information flow secrecy enforcement.

Definition 5.9. The *simple-security property* states that subject s can read an object o only if $SC(s) \geq SC(o)$. Thus, a subject can only read data that at their security class or is less secret. Second, the

★-security property states that subject s can write an object o only if $SC(s) \leq SC(o)$. Thus, a subject can only write data that is at their security class or is more secret.

The simple-security property solves the obvious problem that subjects should not read data that is above their security class. That is, the BLP policy identifies unauthorized subjects for data as subjects whose security class is dominated by the object's security class. Thus, the simple-security property prevents unauthorized subjects from receiving data.

The ★-security property handles the more subtle case that results when the user runs malware, such as a Trojan horse. This property prevents any process from writing secrets to a security class that they dominate, so even if the process is a Trojan horse, it cannot leak data to unauthorized subjects.

The BLP model and its variants are also called multilevel security models and mandatory access control models. A *multilevel security* (MLS) model is a lattice model consisting of multiple sensitivity levels. While the BLP models are simply instances of MLS models, they are MLS models used predominantly in practice. Thus, the BLP models are synonymous with MLS models.

The Bell-LaPadula model implements a *mandatory protection system* (see Definition 2.4 in Chapter 2. First, this model implements a *mandatory protection state*. A fixed set of security classes (labels), consisting of sensitivity levels and categories, are defined by trusted administrators. The dominance relationship among security classes is also fixed at policy definition. Since information flow in the Bell-LaPadula model is determined by the dominance relation, the set of accesses that are possible are fixed.

Second, the Bell-LaPadula model defines a *labeling state* where subjects and objects are labeled based on the label of the process that created them. At create time, a subject or object may be labeled at a security class that dominates the security class of the creating process. Once the subject or object is created and labeled, its label is static.

Third, the Bell-LaPadula model defines a null *transition state*. That is, once a subject or object is labeled (i.e., when it is created), the label may not change. The assumption that the assignment of security classes to subjects and objects does not change is called *tranquility*. Thus, the Bell-LaPadula model satisfies the requirements for a mandatory protection system.

5.3   INFORMATION FLOW INTEGRITY MODELS

Secure operating systems sometimes include policies that explicitly protect the integrity of the system. Integrity protection is more subtle than confidentiality protection, however. The integrity of a system is often described in more informal terms, such as "it behaves as expected." A common practical view of integrity in the security community is: *a process is said to be high integrity if it does not depend on any low integrity inputs.* That is, if the process's code and data originate from known, high integrity sources, then we may assume that the process is running in a high integrity manner (e.g., as we would expect).

Like data leakage, dependence can also be mapped to information flows. In this case, if a high integrity process reads from an object that may be written to by a low integrity process, then the high integrity process may be compromised. For example, if an attacker can modify the configuration files, libraries, or code of a high integrity process, then the attacker can take control of the process, compromising its integrity.

5.3.1   BIBA INTEGRITY MODEL

Based on this view, an information flow model was developed by Biba [27], now called the Biba integrity model [3]. The Biba model is a finite lattice model, as described above, but the model defines properties for enforcing information flow integrity.

Definition 5.10.   The *simple-integrity property* states that subject s can read an object o only if $SC(s) \leq SC(o)$. Thus, a subject can only read data that is at their security class or is higher integrity. Second, the *★-integrity property* states that subject s can write an object o only if $SC(s) \geq SC(o)$. Thus, a subject can only write data that is at their security class or is lower integrity.

Example 5.11.   A Biba lattice model only uses *integrity levels*, not categories. Like the Bell-LaPadula model, the integrity levels are typically totally-ordered. However, unlike the Bell-LaPadula model, there is no commonly-agreed set of levels. An example Biba lattice could include the integrity levels of `trusted`, `system`, `application`, `user`, and `untrusted` where `trusted` is the is highest integrity level and `untrusted` is the lowest. In the Biba model, information flows are only allowed from the higher integrity levels to the lower integrity levels. Thus, subjects and objects that are labeled `untrusted` should not be able to write to subjects or objects in the other levels.

Thus, the flows allowed in a Biba policy are reverse of the flows allowed in the BLP model. Where BLP allows a subject to read objects of a security class dominated by the subject, Biba does not because the objects in a lower security class in Biba are lower integrity. As a result, if we combine BLP and Biba using a single set of security classes, subjects can only read and write data in their security class, which is too limiting for most systems. Also, the integrity and secrecy of a particular object are not necessarily the same, so, in practice two lattices, one for secrecy and one for integrity are created. Further, these lattices contain two distinct sets of security classes. From here, we will call the nodes in a secrecy lattice SC, *secrecy classes*, and the nodes in an integrity lattice SC_i, *integrity classes*. The two sets of security classes are disjoint.

Example 5.12.   Figure 5.4 shows Biba and Bell-LaPadula lattices (levels only) where both must be authorized before an operation is permitted. This joint model enables the enforcement of secrecy and integrity goals. Each subject and object is assigned to a secrecy class and an integrity class which then defines the allowed information flows.

---

[3]There were actually three different integrity models proposed in this paper, called *low-water mark integrity*, *ring integrity*, and *strict integrity*. The strict integrity model became the most prominent of the three models, and it is the one we now call the Biba integrity model. The *low-water mark integrity* gained renewed interest later under the acronym LOMAC, see Section 5.3.2.

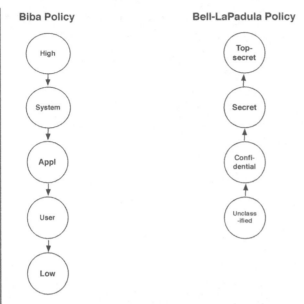

Figure 5.4: For a system that enforces both secrecy and integrity goals, Biba and Bell-LaPadula can be jointly applied. Subjects and objects will be assigned both Biba integrity classes and Bell-LaPadula secrecy classes from the set as shown.

Suppose that an object is a `top-secret, user` object (i.e., secrecy class, then integrity class). Only subjects that are authorized to read both `top-secret` objects according to the Bell-LaPadula policy and `user` objects according to the Biba policy are permitted to read the object. For example, neither `secret, low` nor `top-secret, appl` are allowed to read this object because both the Biba and Bell-LaPadula requirements are not satisfied for these subjects. A subject must be able to both read the object in Bell-LaPadula (i.e., be `top-secret`) and read the object in Biba (i.e., be `low` or `user`).

As for reading, a subject's integrity and secrecy classes must individually permit the subject to write to the object for writes to be authorized.

Lipner defined specific integrity and secrecy levels and categories could be chosen with the aim of constructing a useful composite model for commercial systems [190]. Lipner chose two secrecy levels, *audit manager* and *system low*, where audit manager is more secret (i.e., logs cannot be read once they are written). Also, there are three secrecy categories, *production*, *development*, *system development*. Production is used for production code and data, and the development categories separate system and user program development. In modern systems, we probably need at least one more secrecy level or category for untrusted programs (e.g., to prevent them from downloading the commercial entity's data). For integrity, the lattice has three levels, *system programs*, *operational*, and *system low*. The first two levels separate system code from user applications. There are still two integrity categories, *production* and *development*. Given the wider range of sources of data, more integrity levels are probably necessary today (e.g., Vista defines six integrity levels [152], although it is not a mandatory integrity system).

Biba is derived from Bell-LaPadula in such a way that it is also a mandatory protection system. The Biba protection state is mandatory, and, like Bell-LaPadula, its labeling state only permits the labeling to dominate integrity classes (i.e., lower in the Biba lattice) at creation time. Also, Biba has a null transition state, as integrity class assignments are fixed at creation time.

While the Biba model makes some intuitive sense, it differs from the BLP model in that there are no practical analogues for its use. The BLP model codifies the paper mechanism used by government agencies to protect the secrecy of documents. Thus, there was a practical application of the BLP model, so its application to computerized documents satisfied some users. We note that commercial systems do not use BLP in most cases.

A question is whether equating reading and writing with dependence in the Biba model is a practical way to enforce integrity. Many processes whose integrity must be protected, such as system services, databases, and web servers, accept input from untrusted sources. Biba explicitly prohibits such communication unless a formally-assured (see Chapter 12) *guard* process is inserted to filter such untrusted input. Since such guards must be application-specific, the development of such guards is expensive. Thus, the Biba model has not been applied to the extent that the BLP model has.

5.3.2   LOW-WATER MARK INTEGRITY

An alternative view of integrity is the Low-Water Mark integrity or LOMAC model [27, 101]. LOMAC differs from Biba in that the integrity of a subject or object is set equal to the lowest integrity class input. For example, a subject's integrity starts at the highest integrity class, but as code, libraries, and data are input, its integrity class drops to the lowest class of any of these inputs. Similarly, a file's integrity class is determined by the lowest integrity class of a subject that has written data to the file.

Example 5.13.   Suppose a process *p* uses four inputs whose integrity levels are defined in the integrity lattice of Example 5.12 in Figure 5.4: (1) its code is at the `application` integrity level; (2) its libraries are also at the `system` integrity level; (3) its configuration files are at `application` integrity; and (4) it receives input from untrusted network subjects at the `low` integrity level. As a result, process *p*'s integrity level will be `low`. If process *p* does not receive inputs from untrusted network subjects, then its integrity level will be `application`. However, if it is tricked into using a library that was provided by a untrusted network subject (i.e., the library is at the `low` integrity level), then process *p* will also be `low` protecting the integrity of application data.

LOMAC differs from BLP and Biba in that the integrity class of a subject or object may change as the system runs. That is, a LOMAC transition state is nonnull, as the protection state is not tranquil (see Section 5.2.2). LOMAC relaxes the tranquility requirement securely because it only lowers the security class of a subject or object. For a subject, the lowering of its security class reduces the set of objects that it can modify, reducing the number of objects whose integrity is dependent on the subject. For objects, the lowering of its security class reduces the number of subjects that can read the object, reducing the risk of subjects reading a low integrity object. Tranquility may be relaxed in other ways that still preserve the information flow requirements.

Like Biba, LOMAC does not correspond to how high integrity programs are built in practice, so its use has been limited. Since most high integrity processes receive some low integrity input, a LOMAC policy will result in most processes running at a low integrity class. In practice, systems depend on high integrity programs to protect themselves from malicious input. Despite the fact that there are many instances where this is not the case (e.g., buffer overflows in web server and network daemon software), most applications aim for self-protection.

5.3.3   CLARK-WILSON INTEGRITY

While some secure operating system architectures advocate the extensive use of guards (e.g., MILS, see Chapter 6), it is still an unresolved question whether the expensive of separate guards is justified. In the Clark-Wilson integrity model [54], no such external guard processes are required.

Ten years after the Biba model, Clark and Wilson aimed to bring integrity back into the focus of security enforcement. Clark-Wilson specified that high integrity data, called *constrained data items* (CDIs), must be validated as high integrity by special processes, called *integrity verification procedures* (IVPs), and could only be modified by high integrity processes, called *transformation procedures* (TPs). IVPs ensure that CDIs satisfy some known requirements for integrity (analogously to double-bookkeeping in accounting), so that the system can be sure that it starts with data that meets its integrity requirements. TPs are analogous to high integrity processes in Biba in that only they may modify high integrity data. That is, low integrity processes may not write data of a higher integrity level (i.e., CDI data). These two requirements are defined in two *certification rules* of the model, CR1 and CR2.

The Clark-Wilson model also includes *enforcement rules* that limit the users and TPs that may access CDIs, ER1 and ER2. When a CDI is accessed, it can be accessed only using a TP authorized for that CDI (ER1), and only by a user authorized to run that TP to access that CDI (ER2).

The Clark-Wilson model is comprehensive in that it defines rules for authentication, auditing, and administration. Clark-Wilson requires that all users must be authenticated before they can run a TP (ER3). Also, auditing is enforced by the rule that states that all TPs must append operational information sufficient to reconstruct any operation in an append-only CDI (CR4). Administration is enforced via two rules. First, Clark-Wilson restricts the certifier of a TP to be an entity who does not have execute permission for that TP (ER4). Second, administration of permissions assigning users and TPs to CDIs must satisfy the principle of *separation of duty*, defined by Clark and Wilson. According to this principle, no single principal can execute all the transactions in a separation transaction set. For example, a subject cannot both execute a payment and authorize that payment.

However, the most significant part of the model was the portion concerned with integrity protection. The Clark-Wilson model also includes a rule (CR5) to prevent the high integrity processes (IVPs and TPs) from depending on low integrity data. A high integrity process may read low integrity data, called *unconstrained data items* (UDIs), but it must either *upgrade or discard* that data upon receipt. That is, Clark-Wilson does not require a separate guard process like Biba, but it requires that the high integrity process guard itself. In order for a high integrity process to justify its ability to both write CDIs correctly and protect itself when it reads UDIs, the Clark-Wilson model requires that TPs be fully assured themselves. As we have discuss in Chapter 12, methodologies for comprehensive assurance are expensive, so few applications have ever been assured at this level.

5.3.4   THE CHALLENGE OF TRUSTED PROCESSES

With MLS and Biba, we can formally show that a system's information flows adhere to these policies, but they assume that there are no processes that would ever require illegal information flows. For MLS, some processes may be required to leak some information processed at a secret clearance to processes that can only read public information. This is analogous to a general holding a news conference or submitting orders to subordinates. For Biba, some trusted processes may be required to process inputs received from untrusted processes. This is analogous to an oracle providing responses to questions. The integrity of the oracle cannot be impacted by the integrity of the questions.

Such computations must be implemented by trusted processes. In Multics, it was found [333] that system processes could "not operate at a single clearance level," so these processes must be trusted "never [to] violate the fundamental security rules." In the design of KSOS, the set of trusted user-level programs including 10 different groups of programs [97].

While MLS and Biba models support the use of trusted processes (e.g., guards), the number of programs that need to be trusted became nontrivial and remain so today. For example, SELinux/MLS (see Chapter 9) specifies over 30 trusted subjects. One approach has been to reduce the scope of trust required of such programs. GEMSOS models permitted processes to work within ranges of access classes [277, 290]. The Caernarvon model [278] permits processes to work within access class ranges encompassing both secrecy and integrity. To obtain such authorization, programs must be certified in some manner analogous to Common Criteria assurance.

In lieu of formal assurance models, other approaches for building and using trusted processes have emerged. For example, *security-typed languages* ensure that compiled programs satisfy associated secrecy and integrity requirements [291, 219]. Such languages depend on trust in compilers and runtime environments that are not presently assured. Nonetheless, the idea of leveraging language level guarantees is gaining momentum.

Presuming that a basis for trust in programs emerges, a variety of recent work is already exploring how to manage such trust. A number of researchers have reverted to information flow as the ideal basis for security, and the problem is now to define the parameters of trust over information flow.

Some recent models leverage the concepts of the *decentralized label model* (DLM) [218] used in security-typed languages at the system-level (i.e., per process) to ensure information flow and define trust in processes. The key features of the DLM model are its representation of both secrecy and integrity in information flow policies, and its inclusion of specifications of trust in downgrading secret information (*declassifiers*) and upgrading low integrity information (*endorsers*). The Asbestos label model [316] expresses both secrecy and integrity requirements, and provides a special level value ⋆ that designates trust, but this trust can be limited to specific categories of data. Subsequent generalization of the Asbestos model into the *decentralized information flow control* (DIFC) model even more closely resembles the DLM model [174, 349, 350]. Processes may change the label of data under the limitations of information flow, unless the process has a privilege that permits it to downgrade or upgrade per specific labels. For example, a privilege t^- permits a holding process to remove the label t from its data, thus permitting it to downgrade such data.

Other recent integrity models define how processes may manage low integrity data, yet be trusted to maintain their high integrity [182, 285, 300]. For example, CW-Lite [285] leverages the semantics of the Clark-Wilson where a high integrity process may receive low integrity data if it immediately discarded or upgrades such data. In CW-Lite, trusted, filtering interfaces are identified as the only locations that such discard/upgrade is permitted for a process. Only such interfaces must be trusted. Usable Mandatory Integrity Protection (UMIP) [182] has a similar approach, but it defines trust in terms of the types of information flows (e.g., network, IPC) that a process may be entrusted. Finally, Practical Proactive Integrity (PPI) [300] permits definition of all previous integrity model semantics into policy options that enable flexible definition of trust in a process's handling of low integrity data.

5.4 COVERT CHANNELS

Lampson identified the problem that systems contain a variety of implicit communication channels enabled by access to shared physical resources [177, 189]. For example, if two processes share access to a disk device, they can communicate by testing the state of the device (i.e., whether it is full or not). These channels are now called *covert channels* because they are not traditionally intended for communication. Such channels are present in most system hardware (e.g., keyboards [284], disks [160], etc.) and across the network (e.g., [110, 294, 338]). Millen has provided a summary of covert channels in multilevel security systems [212].

From a security perspective, the problem is that these communication mechanisms are outside the control of the reference monitor. Assume a BLP policy. When a secret process writes to a secret file, this is permitted by BLP. When a unclassified process writes to an unclassified file this is also permitted. However, if the two files are stored on the same disk device, when the secret process fills the disk (e.g., via a Trojan horse) the unclassified process can see this. Thus, a covert communication is possible. Different covert channels have different data rates, so the capacity of the channel is an important issue in determining the vulnerability of the system [211, 215].

5.4.1   CHANNEL TYPES

Covert channels are classified into two types: storage and timing channels. A *storage covert channel* requires the two communicating processes to have access to a shared physical resource that may be modified by the sending party and viewed by the receiving party. For example, the shared disk device is an example of a storage channel because a process may modify the disk device by adding data to the disk. This action is observable by another party with access to this disk because the disk contents may be consumed. For example, a full disk may signal the binary value of one, and an available disk may signal the binary value of zero. Uses of the disk as a storage channel have been identified [160]. Additionally, the communicating parties also need a synchronized clock in order to know when to test the disk. For example, how do we differentiate between the transmission of a single "one" and two consecutive "ones." Of course, a system has many forms of time keeping available, such as dates, cycle counters, etc. As a result, storage covert channels are quite practical for the attacker [315].

A *timing covert channel* requires that the communicating party be able to affect the timing behavior of a resource [341]. For example, if a communicating process has high priority to a network device, then it can communicate by using the device for a certain among of time. For example, if the communicating process is using the device, then it is transmitting a binary "one," otherwise it is transmitting a binary "zero." Reliable timing channels are a bit harder to find because often there are many processes that may affect the timing of a device, and the scheduling of the device may not always be so advantageous to communication. In addition to the source whose timing behavior may be affected, timing channels also require synchronized clock.

Covert channels are problematic because they may be built into a system unknowingly, and they may be difficult to eradicate completely. A storage channel is created whenever a shared resource of limited size is introduced to the system. This may be a storage device, or it may be artificially created. For example, limiting the number of sockets that a system may have open at one time would introduce a storage channel. The challenge is to identify the presence of such channels. Once they are found, they are easily removed by prohibiting the processes from using the same resource.

Initially, the techniques to identify covert channels in software was ad hoc, but researchers developed systematic techniques to enable the discovery of covert channels. First, Kemmerer defined the Shared Resource Matrix [162, 163] (SRM) where shared resources and the operations that may access these resources are identified. A matrix is then constructed that shows how the resources are accessed by the operations. Given the matrix, we can determine whether a high process can use the resource and operations to leak data to the low process. The SRM technique is a manual technique for reviewing source code, but later techniques analyzed source code directly [40, 164]. For example, Kemmerer later defined Covert Flow Trees [164] (CFT) in which trees represent the flow of information in a program that serves shared resources, such as a file system. See Bishop [29] for a more detailed description of these approaches as well.

While it is at least theoretically possible to remove all covert storage channels from a system, timing channels cannot be completely removed. This is because the timing behavior of a system is available to all processes. For example, a process can cause a longer page fault handling time by consuming a large amount of pages [4]. Techniques to address timing channels, such as *fuzzy time* [140, 311], involve reducing the bandwidth of channels by randomizing their timing behavior. In a fuzzy time system, the response time of an operation is modified to prevent the operation's process from communicating. In theory, the time of every operation must be the same in order to prevent any communication. Thus, operations must be delayed in order to hide their performance (i.e., slow operations cannot be made faster), thus effecting the overall performance of the system. Other techniques to address timing channels have similar negative performance effects [160].

A question is whether BLP enforcement could be extended to control covert channels. If an *unclassified* process stores files on a disk, then the disk must be assigned the *unclassified* access class because the unclassified process can "read" the state of the disk. Thus, we would not store secret files on the disk, so a Trojan horse running in a *secret* process could not write to this disk. Unfortunately, this analogy cannot be carried out completely, as some resources, such as CPU, must be shared, and the cost of having a device per access class is sometimes impractical (e.g., one network device per access class). As a result, the covert channel problem is considered as an implementation issue that requires analysis of the system [249].

5.4.2   NONINTERFERENCE

An alternative for controlling covert channels is to use models that express the input-output requirements of a system. These models are based on the notion of *noninterference* [113]. Intuitively, noninterference among processes requires that the actions of any process have no effect on what any other process sees. For example, the actions of one process writing to a file should not be seen by any other process should the system enforce noninterference between them.

More formally, noninterference has been assessed as follows [113, 202, 203]. Consider a system in which the output of user u is given by the function $out(u, hist.read(u))$ where $hist.read(u)$ is the trace of inputs to u and $read(u)$ was the latest input. Noninterference is defined based on what part of a trace can be purged from the input of other users u' (and their processes) whose security class $SC(u')$ is dominated by $SC(u)$.

Definition 5.14. Let *purge* be a function from *users* × *traces* to *traces* where $purge(u', hist.command(u)) = purge(u'.hist)$ if $SC(u') < SC(u)$.

That is, *purge* ensures that lower secrecy subjects are not impacted by the commands run by higher secrecy subjects.

In general, noninterference is not comparable to BLP. Since BLP does not prevent covert channels, BLP enforcement is weaker than nonintereference. However, noninterference does not prevent the inputs of lower secrecy subjects from impacting higher secrecy subjects, implying that lower secrecy subjects may learn the state of higher secrecy data (i.e., illegally read up). In addition,

---

[4]This attack has a combination of storage and timing issues. The size of memory is finite which implies a storage channel, but the measurement of value involve the timing of the memory access.

noninterference traces do not address problems caused by timing channels, as such channels are not represented in the traces. On the other hand, noninterference prevents the more secret subject from passing encrypted data to the secret subject [302], so noninterference, like the other security properties that we have discussed, is also a more conservative approximation of the security than necessary, in practice. There is a large body of formal modeling of noninterference and assessment of its properties [70, 203, 218, 261, 144, 55, 91, 246], but this beyond the scope of this book. In practice, nonintereference properties are too complex for manual analysis, and while research applications have had some successes [33, 125], support for enforcing noninterference requirements, in general, has not been developed.

5.5 SUMMARY

In this chapter, we examined the policy models that have been used in mandatory protection systems. Such policies must provide mandatory protection states that define the desired secrecy and integrity protection required, labeling states that securely assign system processes and resources to security classes, and transition states that ensure that any change in a process or resource's security class is also secure. These models define security goals, labeling, and transitions in terms of information flow. Information flow describes how any data in the system could flow from one subject to another. Information flow is conservative because it shows the possible flow paths, but these may or may not actually be used. The information flow abstraction works reasonably well for secrecy, at least for the government sector where a version of information flow was employed prior to computerization. However, information flow has not been practical for integrity, as the abstraction is too conservative.

Secrecy is defined in terms of a finite lattice of security classes where the security class determines who is authorized to access the data. It turns out that the more secret the data, the fewer the subjects that can read it. This has been codified in the Bell-LaPadula (BLP) model. Only subjects whose security class dominates or is equal to that of the data may read it (*simple-security property*). However, in order to prevent leakage via malware, subjects can only write to objects whose security class dominates that of the subject (*★-security property*).

For integrity, the information flow requirements are reversed. This is because we are concerned about high integrity processes reading less trusted data. Thus, the *simple-integrity* and *★-integrity* properties are defined with reverse flows. The Biba model defines these requirements. Since it was found that integrity was largely independent of secrecy, a second set of security classes is generally used. In practice, information flow has been too conservative an approximation for integrity because data input does not imply *dependence*. Few secure operating systems use formal integrity management at present, despite the many vulnerabilities that result from the improper handling of malicious input.

Finally, we examined the problem of covert information flows. In these, access to shared resources is used to convey information. If a subject can modify a shared resource, the others that use this resource can receive this "signal." In theory, we cannot prevent two processes running on the same system from using a covert channel availed by the system, but we can reduce the bitrate. The *noninterference* property formalizes information flow in terms of input and output, such that a covert channel must convey some input that affects the behavior of an unauthorized subject. However, practical application of noninterference is complex and supporting tools are not available.

CHAPTER 6

Security Kernels

While the Multics project was winding down in the mid-1970s, a number of vendors and researchers gained confidence that a secure operating system could be constructed and that there was a market for such an operating system, within the US government anyway. Many of the leaders of these operating system projects were former members of the Multics team, but they now led other research groups or development groups. Even Honeywell, the owner of the Multics system, was looking for other ways to leverage the knowledge that it gained through the Multics experience.

While the Multics security mechanisms far exceeded those of the commercial operating systems of the day, it had become a complex system and some of the decisions that went into its design needed to be revisited. Multics was designed to be a general-purpose operating system that enforced security goals, but it was becoming increasingly clear that balancing generality, security, and performance was a very difficult challenge, particularly given the performance of hardware in the mid-70s. As a result, two directions emerged, one that focused on generality and performance with limited security mechanisms (e.g., UNIX) and another that emphasized verifiable security with reasonable performance for limited application suites (i.e., the security kernel). In the former case, popular, but insecure, systems (see Chapter 4) were built and a variety of efforts have been subsequently made to retrofit a secure infrastructure for such systems (see Chapters 7 through 9). In this chapter we examine the latter approach.

In the late 1970s and early 1980s, there were several projects that aimed to build a secure operating system from scratch, addressing security limitations of the Multics system. These included the Secure Communications Processor (Scomp) [99] from Honeywell, the Gemini Secure Operating System (GSOS or GEMSOS) [290] from Gemini, the Secure Ada Target (SAT) [34, 125, 124] and subsequent LOCK systems [293, 273, 274, 292, 276] from Honeywell and Secure Computing, respectively, which are based on the Provably Secure Operating System (PSOS) design [92, 226], the Kernelized Secure Operating System (KSOS) [198] from Ford Aerospace and Communications, the Boeing Secure LAN [298], and several custom guard systems (mostly proprietary, unpublished systems). In this chapter, we examine two of these systems, Scomp and GEMSOS, to demonstrate the design and implementation decisions behind the development of security kernels. These two systems represent two different implementation platforms for building a security kernel: Scomp uses custom hardware designed for security enforcement, whereas GEMSOS was limited to existing, commercially-popular hardware (i.e., the Intel x86 platform). These systems show what can be done when even the hardware is optimized for security (Scomp) and the limitations imposed on the design when available hardware is used (GEMSOS). Recent advances in commercial hardware, such as I/O MMUs, may enable us to revisit some Scomp design decisions.

6.1 THE SECURITY KERNEL

The major technical insight that emerged at this time was that a secure operating system needed a small, verifiably correct foundation upon which the security of the system can be derived. This foundation was called a *security kernel* [108]. A security kernel is defined as the hardware and software necessary to realize the reference monitor abstraction [10]. A security kernel design includes hardware mechanisms leveraged by a minimal, software trusted computing base (TCB) to achieve the reference monitor concept guarantees of tamperproofing, complete mediation, and verifiability (see Definition 2.6).

The first security kernel was prototyped by MITRE in 1974. It directly managed the system's physical resources with less than 20 subroutines in less than 1000 source lines of code. In addition to identifying *what* is necessary to build a security kernel that implements a reference monitor, this experience and the Multics experience indicated *how* a security kernel should be built. While mediation and tamperproofing are fundamental to the design of a security kernel, in building a security kernel the focus became verification. Three core principles emerged [10]. First, a security kernel has to *implement a specific security policy*, as it can only be verified as being secure with respect to some specific security goals. A security goal policy (e.g., based on information flow, see Chapter 5) must be defined in a mandatory protection system (see Definition 2.4) to enable verification. Second, the design of the security kernel must define a verifiable *protection behavior of the system as a whole*. That is, the system mechanisms must be comprehensively assessed to verify that they implement the desired security goals. This must be in the context of the security kernel's specified security policy. Third, the implementation of the kernel must be shown to be *faithful to the security model's design*. While a mathematical formalism may describe the design of the security kernel and enable its formal verification, the implementation of the security kernel in source code must not invalidate the principles established in the design.

Thus, the design and implementation of security kernels focused on the design of hardware, a minimal kernel, and supporting trusted services that could be verified to implement a specific security policy, multilevel security. While Multics had been designed to implement security on a particular hardware platform, the design of security kernels included the design of hardware that would enable efficient mediation of all accesses. The design of security kernel operating systems leverages this hardware to provide a small number of mechanism necessary to enforce multilevel security. Finally, some trusted services are identified, such as file systems and process management, that are necessary to build a functional system.

The primary goal of most security kernel efforts became *verification* that the source level implementation satisfies the reference monitor concept. This motivated the exploration of formal and semi-formal methods for verifying that a design implemented the intended security goals and for verifying that a resultant source code implementation satisfied the verified design. As Turing showed that no general algorithm can show that any program satisfies a particular property (e.g., halts or is secure), such security verification must be customized to the individual systems and designs. The work in security kernel verification motivated the subsequent methodologies for system security assurance (see Chapter 12). The optimistic hope that formal tools would be developed that could automatically support formal assurance has not been fulfilled, but nonetheless assurance is still the most practical means known to ensure that a system implements a security goal.

Verification that an implementation correctly enforces a system's security goals goes far beyond verifying the authorization mechanisms are implemented correctly. The system implementation must be verified to ensure that all system resource mechanisms (see Chapter 1) are not vulnerable to attack. As computing hardware is complex, assurance of correct use of hardware for implementing system resources is nontrivial. Consider the memory system. A hardware component called the Translation Lookaside Buffer (TLB) holds a cache of mappings from virtual memory pages to their physical memory counterparts. If an attacker can modify these mappings they may be able to compromise the secrecy and integrity of system and user application data. Depending on the system architecture, TLBs may be filled by hardware or software. For hardware-filled TLBs, the system implementation must ensure that the page table entries used to fill the TLB cannot be modified by attackers. For software-filled TLBs, the refill code and data used by the code must be isolated from any attacker behavior. Further, other attacks may be possible if an attacker can gain access to secret memory after it is released. For example, heap allocation mechanisms must be verified to ensure clearing of all secret memory (e.g., to prevent *object reuse*). Even across reboots, secret data may be leaked as BIOS systems are inconsistent about whether they clear memory on boot or not, and data remains in memory for sometime after shutdown. As a result of these and other possible attack vectors (e.g., covert channels, see Chapter 5), careful verification of system implementations is necessary to ensure reference monitor guarantees, but it is a complex task.

In this chapter, we examine two of systems whose designs aimed for the most comprehensively assured security, Honeywell's Scomp [99] and Gemini's GEMSOS [290]. Both these systems achieved the highest assurance rating ever achieved for an operating system, A1 as defined by the Orange Book [304] assurance methodology [1]. Scomp was used as the basis for the design of the assurance criteria. GEMSOS is still available today [5].

6.2   SECURE COMMUNICATIONS PROCESSOR

The Honeywell Secure Communications Processor (Scomp) system is a security kernel-based system [99] designed to implement the Multics's multilevel security (MLS) requirements [23], see Chapter 3. The original idea was to build a security kernel and an emulator to enable execution of an ordinary operating system (UNIX), as was done by KSOS [198] and the UCLA Secure Data UNIX system [248]. After the performance and security of such emulated systems was found to be insufficient, a decision was made to build a new application interface for Scomp that provides applications with the necessary security that runs with reasonable performance.

The performance of a emulated system run on a security kernel is impacted by two issues. First, the emulation may involve converting between incompatible representations of the two systems. For example, UNIX I/O copies data directly to the application's address space (e.g., on a file or

<hr>

[1]The GEMSOS A1 evaluation was as part of the BLACKER system.

network read), but Scomp maintains data in individually managed segments to which access must be authorized. As a result, rather than getting a filled buffer as in UNIX, Scomp I/O provides a reference to a segment with the data. Second, the hardware features of a system may not supply efficient primitives for emulated function. For example, Multics hardware did not provide hardware support for ring crossings (i.e., protection domain transitions), so these must be implemented in software at a higher cost.

Further, the emulated system may include mechanisms that are not secure with respect the requirements of the security kernel. For example, UNIX supports the transfer of file descriptors on `fork` and `exec` operations. Thus, a parent process may be able to leak data to the child or provide the means for the child to leak its own data. This problem must be addressed in secure versions of commercial operating systems, see Chapter 7, but the Scomp designers felt these and similar issues warranted a new application interface. The conflict between functional interfaces and how to secure them is fundamental to the design of secure systems.

As a result of these performance and security concerns, the Scomp designers developed not only a security kernel, but also new hardware mechanisms and a new application interface for writing programs to the security kernel. Below, we discuss the overall architecture, major features, and impact on application development.

6.2.1   SCOMP ARCHITECTURE

The Scomp system architecture is shown in Figure 6.1. The *Scomp trusted computing* base consists of three components running in rings 0, 1, and 2. The Scomp Trusted Operating System (STOP) consists of a security kernel running in ring 0 and trusted software running in ring 1. The trusted functions of the Scomp Kernel Interface Package (SKIP) run in ring 2. Applications access protected resources managed by the Scomp trusted computing base using the SKIP library running in the application's address space.

The Scomp security kernel mediates access to all protected resources using an MLS policy. When an application process needs access to a protected resource (i.e., a memory or I/O segment), it must ask the security kernel for a *hardware descriptor* sufficient to access this resource. The security kernel authorizes whether the process can access the resource, and if authorized, builds a hardware descriptor for accessing this resource. The security kernel stores the hardware descriptor and returns a reference to the descriptor to the process for subsequent use. A Scomp hardware descriptor includes an object reference and the authorized access permissions for that process.

Isolation, and hence tamperproofing, is implemented by a ring protection mechanism. Similar to Multics, an access bracket mechanism controls whether code in one ring is permitted to request services from another ring. Unlike Multics, all rings and ring transitions are implemented in hardware.

Complete mediation is implemented in hardware. All requests for memory or device access are mediated by security protection hardware described below in Section 6.2.2.

For the first time in an operating system development process, verification that the Scomp security model and implementation enforce the MLS policy was a first-class task. Scomp's trusted

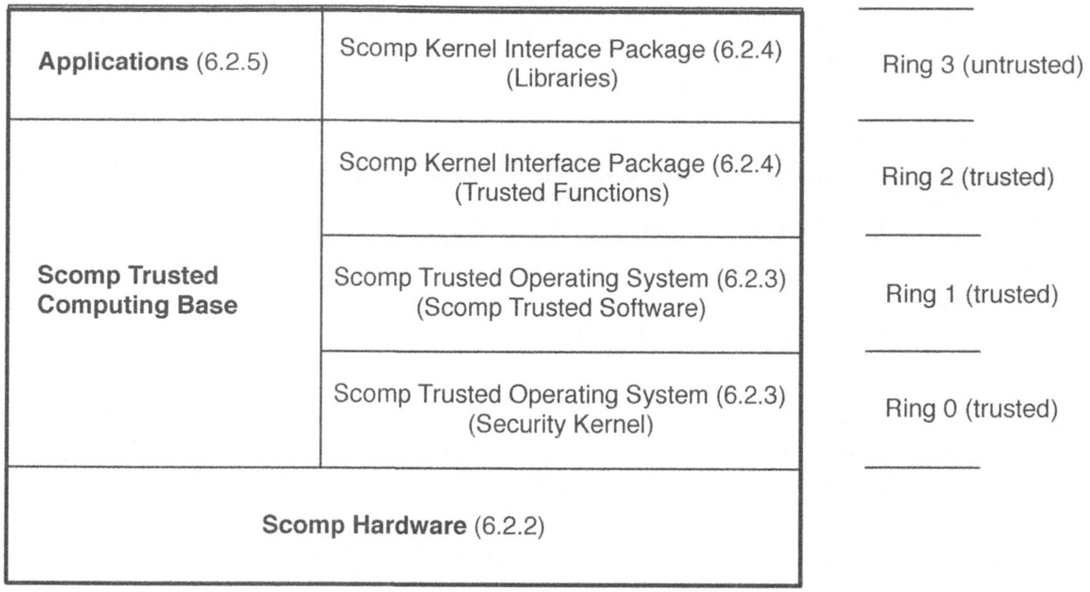

Applications (6.2.5)	Scomp Kernel Interface Package (6.2.4) (Libraries)	Ring 3 (untrusted)
Scomp Trusted Computing Base	Scomp Kernel Interface Package (6.2.4) (Trusted Functions)	Ring 2 (trusted)
	Scomp Trusted Operating System (6.2.3) (Scomp Trusted Software)	Ring 1 (trusted)
	Scomp Trusted Operating System (6.2.3) (Security Kernel)	Ring 0 (trusted)
Scomp Hardware (6.2.2)		

Figure 6.1: The Scomp system architecture consists of hardware security mechanisms, the Scomp Trusted Operating System (STOP), and the Scomp Kernel Interface Package (SKIP). The Scomp trusted computing base consists of code in rings 0 to 2, so the SKIP libraries are not trusted.

software is verified using two technologies. First, SRI's Hierarchical Development Methodology [48] is used to verify that a formal model of the security kernel's specification, called the *formal top-level specification* (FLTS), enforces the MLS policy. Second, trusted software outside the kernel is verified using a procedural specification applied using the Gypsy methodology [118].

Based on the hardware enforcement features (complete mediation), protected software in the trusted computing base (tamperproofing), and the formal verification of the security kernel and other trusted software (verification), Scomp defined a process for building secure systems to satisfy the reference monitor concept. This process became the basis for the A1 evaluation level (i.e., the most secure evaluation level) of the DoD's Trusted Computer System Evaluation Criteria [304].

6.2.2 SCOMP HARDWARE

The Scomp hardware design is based on the ideas of the Multics system with two key changes. First, the Multics protection ring mechanism is concentrated in four protection rings and is extended to enforce more limited access for applications on ring transitions. Second, a *security protection module* is defined to mediate all memory and I/O accesses in the Scomp system.

First, the Scomp hardware implements four protection rings. The security kernel runs in the most privileged protection ring, ring 0, and user software runs in the least privileged ring, ring 3. Trusted software outside of the security kernel may occupy either ring 1 or 2.

The Scomp hardware supports a call-return mechanism that enables procedures in a less-privileged ring to invoke a procedure in a higher-privileged ring. The Scomp call-return mechanism is similar to the ring bracket access mechanism in Multics. However, Scomp hardware also provides a mechanism to access caller-supplied arguments at the caller's privilege level, called *argument addressing mode*. This mechanism enables the kernel to prevent itself from accessing data that the caller could not access. For example, the kernel interface can define a memory-mapping operation that requires the caller to supply a memory page. Using argument addressing mode, the kernel could only use the memory reference if the caller has access to this memory reference. This prevents the *confused deputy problem* [129] in a manner analogous to capability systems, see Chapter 10.

Second, the Scomp hardware includes a component called the *security protection module* (SPM) that provides a tamperproof service to mediate all memory and I/O accesses, as required for a reference monitor. As shown in Figure 6.2, the SPM mediates the main system bus (called the Level 6/DPS 6 bus) that provides access to peripherals and memory (e.g., memory and PCI buses combined). Also, the virtual memory interface unit uses the SPM to convert virtual address to physical segment addresses.

Each process has a *descriptor base root* that references the memory and I/O descriptors available to a process. Any virtual memory access references a descriptor that is used to authorize the request and access the physical location. Memory descriptors contain a pointer to physical memory, access permissions, and memory management data. Note that each access to words in memory is mediated by the SPM, so an access check accompanies any reference to memory. Since each instruction may make a memory reference for its instruction and its operands, plus access to page table entries, multiple access control checks may occur on each instruction. Accesses that hit in the hardware cache do not incur a memory reference and its accompanying authorization. Contrast this with modern systems, which authorize memory access at the page level, with a simple check on the page table or TLB entry for read or write access.

Mediation of I/O is similar to that for memory. An I/O operation is a request to a virtual name for a physical device. The SPM uses this virtual name to retrieve an I/O descriptor for the device. Access to this I/O descriptor is authorized similarly to memory descriptors. This mechanism supports two types of *direct memory access* (DMA) [2]. First, for premapped DMA, the SPM authorizes access to the device and the memory prior to the first request only. Since the I/O descriptors have been authorized and are used for each I/O request, subsequent requests no longer require authorization (i.e., the authorized descriptors are cached). Second, for mapped I/O, the I/O addresses for DMA are sent to the device. Thus, each access of the device to physical memory is authorized by the SPM (e.g., based on the process that owns the device).

Because the kernel builds the I/O descriptor and the hardware authorizes the I/O operation, it is possible to run I/O commands (i.e., drivers) in unprivileged processes (i.e., outside the kernel). This has performance and security advantages. First, once the access is mediated, the user process may interact with the devices directly, thus removing the need for kernel processing and context

---

[2]Direct memory access (DMA) enables devices to write into the system's physical memory without involving the system's CPU.

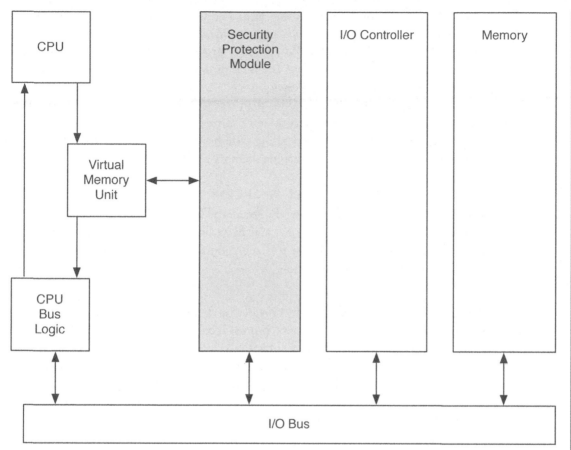

Figure 6.2: The Scomp *security protection module* (SPM) mediates all accesses to I/O controllers and memory by mediating the I/O bus. The SPM also translates virtual addresses to physical segment addresses for authorization.

switches. Second, drivers have been shown to be the source of many kernel errors [82], so the ability to remove them from the trusted computing base would improve software correctness. Further, in this architecture, new I/O devices can be added without modifying the kernel or the SPM, so the system's extensibility does not suffer from this approach.

Modern hardware, such as the x86 architecture, adopted the four-ring architecture of Scomp, but not the I/O mediation provided by the SPM. As a result, such systems are vulnerable to DMA devices. Buggy or malicious drivers can configure a DMA device to write at any physical memory location, so kernel memory may be overwritten. Since DMA bypasses the CPU, the kernel cannot prevent such writes in software. Recently, both Intel and AMD have released processors with I/O memory protection, called an I/O MMU [141, 8]. An I/O MMU also mediates an I/O bus (the

PCI bus), but the devices communicate using virtual addresses rather than I/O descriptors. The I/O MMU translates these *virtual I/O addresses* to physical addresses, authorizing access to the resultant physical address. The I/O MMU approach provides the advantages envisioned by the SPM architecture. For example, *passthrough I/O* built on I/O MMUs enables I/O to be conveyed directly from the device to untrusted processes.

The use of virtual I/O addresses rather than the I/O descriptors as in Scomp reflects the change from segmented protection of early systems to page protection of modern systems. While segmentation is still supported, modern operating systems use address spaces based on sets of pages managed by *page tables* rather than by managing memory segments.

6.2.3   SCOMP TRUSTED OPERATING PROGRAM

Technically, the Scomp Trusted Operating Program (STOP) consists of three components: (1) a *security kernel*; (2) a set of *trusted software*; and (3) a Scomp kernel interface package for user applications. We describe the first two in this subsection, and discuss user applications and the interface package in the following subsection.

Security Kernel   The Scomp security kernel provides fundamental system processing in ring 0. The security kernel provides memory management, process scheduling, interrupt management, auditing, and reference monitoring functions. Consistent with the idea of a security kernel, the function of the Scomp security kernel is minimized to reduce the amount of trusted code. As a result, the Scomp security kernel is only 10K source lines of code, mostly written in Pascal.

Kernel objects consist of processes, segments, and devices. Each are identified by a globally-unique, immutable 64-bit identifier. The kernel maintains access control information and status data for each object. The Scomp access control model is essentially the same as the Multics approach, see Chapter 3, consisting of multilevel security (i.e., Bell-LaPadula sensitivity levels and category sets [23]), ring bracket policies, and discretionary policies. The ring bracket representation is modified to describe access based on the owner of the object, groups, or others. Thus, an owner may be allowed to access an object from a different set of rings than an arbitrary user.

The security kernel defines 38 *gates* for processes running outside the kernel to invoke kernel services. Gates are analogous to system calls in modern operating systems, providing function to create objects, map segments, and pin physical memory to virtual addresses. Scomp gates are also analogous to Multics gates in that they provide argument validation to protect the integrity of security kernel.

Trusted Software   Scomp trusted software runs services that do not require ring 0 privilege, but provide functions that must be trusted to enforce control on user applications properly. There are two types of trusted software. The first type of trusted software is trusted not to violate system secrecy or integrity goals. For example, secure loader is trusted to load a process for any subject that ensures correct enforcement of that subject's information flows. The second type of trusted software is trusted to maintain the security policy correctly. For example, services that modify user authentication data must be trusted.

Scomp has 23 processes that implement trusted functions, consisting of 11K source lines of code, written in the C language. There are three general types of trusted software. First, *trusted user services* provide an interface to Scomp for the user. User services include login, discretionary access control management, mandatory level selection, and process management. Second, *trusted operation services* provide functions that enable the system administrators to manage the system. Scomp require system management by operators who started the system, maintain mandatory policies (multilevel security and ring bracket policies), and collect and evaluate audit logs. Trusted operation services include a secure startup service and various operator commands (e.g.,setting time). Third, Scomp *trusted maintenance services* enable the system administrator to modify system data, such as install new versions of programs.

Scomp trusted software are invoked via a trusted communications path with the user. This is the familiar "secure attention sequence" used in modern systems. The purpose of this communication path is to prevent malicious software (e.g., Trojan horses) from masquerading as the user (e.g., trying to guess passwords and login). Further, the user also learns that she is communicating directly with the trusted software when the trusted communications path is invoked. Only the kernel can receive the interrupt invoked, so the user can be certain that the resulting response originated from trusted software.

6.2.4   SCOMP KERNEL INTERFACE PACKAGE

The Scomp kernel interface package (SKIP) provides a uniform interface for user applications to access trusted functions. SKIP code is divided into two parts. First, SKIP functions implement trusted operations on user-level objects: files via a hierarchical file system, processes via process management, and concurrent I/O via an event mechanism. SKIP functions are allowed to manipulate system state, so they are also trusted not to violate MLS requirements, like trusted software. Second, a SKIP library provides a high-level interface for accessing such functions. SKIP libraries run in the protection domain of user applications, so they are not trusted with system state. SKIP functions run in ring 2, and the libraries run with user applications in ring 3.

SKIP functions in ring 2 are also invoked via gates, similarly to the kernel and trusted software. For example, calls to modify the file system state, such as renaming a file, are invoked from a user application using the SKIP library which invokes a SKIP gate before being processed in ring 2. Thus, file system operations are protected from user applications.

SKIP provides a library for user applications to access files, modify file contents, and manage the file hierarchy. The actual file system operations, and its state, are maintained by the SKIP functions in ring 2, in the manner described above. All file operations are authorized based on the requestor's sensitivity level and ring number. This results in a file system hierarchy where the sensitivity level is nondecreasing from the root.

The SKIP library also enables applications to perform various kinds of I/O. In effect, the device drivers are provided in the SKIP library. As described above, the SPM provides a protection mechanism that mediates application access to devices and device access to memory segments. To enable concurrency control across multiple applications, the SKIP function provides an event mechanism in ring 2. The event mechanism processes interrupts, maintains a queue of requests, and provides event notification. Handlers may be defined by user applications, so they are run in the library.

6.2.5   SCOMP APPLICATIONS

The Scomp hardware and software is general purpose, but it defines a completely new application interface, so new application software needs to be constructed. Actually, the original idea was to run a UNIX emulator on Scomp, so UNIX applications could be run, but emulation was found to be too slow and too insecure as described above. There was also a plan to map UNIX system calls to the SKIP interface. The results of this effort do not appear to have been documented.

For example, the Scomp system was the basis for a *mail guard* [71]. Multilevel security ensures that secrets are protected from leakage, but in a military environment orders, often based on the analysis of secret information, must be conveyed to less cleared subordinates. In order to ensure that orders are delivered without leaking secrets, assured mail guards are run on the Scomp environment. The mail guards evaluate the content of the orders using specialized filters before forwarding them. Scomp is an ideal platform as it is assured to enforce the multilevel security requirements and execute the mail guard without allowing malicious software to take control of its execution.

Other general purpose applications used in an multilevel secure environment may also benefit from Scomp, so Honeywell developed other applications as well. For example, Honeywell implemented the Secure Office Management System for Scomp [100]. The software consists of a word processor, email, spreadsheet, database, and printing support. Such software provides a variety of features with each function, and is multilevel-aware to help the user navigate issues with information secrecy management and release. The challenge is that other vendors developed office software for ordinary operating systems with greater features, and these became the de facto standard for office processing. As a result, the government employees adopted this software and the insecure environments in which they run.

6.2.6   SCOMP EVALUATION

1. **Complete Mediation**: How does the reference monitor interface ensure that all security-sensitive operations are mediated correctly?

   Scomp performs all mediation in hardware, so complete mediation is always performed on the correct system resource.

2. **Complete Mediation**: Does the reference monitor interface mediate security-sensitive operations on all system resources?

All system resources are segments, memory and I/O, and all instructions access segments, so the hardware-based mediation of Scomp mediates all security-sensitive accesses.

The Scomp file system in ring 2 controls access to files, so higher-level, file policies may be written. However, initial access to file data depends on access to I/O. The file system must be trusted to prevent unauthorized access to one process' file data by another process.

3. **Complete Mediation**: How do we verify that the reference monitor interface provides complete mediation?

   Hardware verification justifies complete mediation in Scomp.

4. **Tamperproof**: How does the system protect the reference monitor, including its protection system, from modification?

   Scomp uses a protection rings to protect the security kernel from unauthorized modification. The security kernel runs in ring 0, and only 38 gates permit access to the kernel from other protection rings.

   Scomp uses a more complex version of the Multics discretionary ring bracket integrity model to express access to ring 0. Since it must be possible to update the kernel (e.g., by updating the file system), there are some subjects (and their processes) that could modify the kernel, including the protection system and reference monitor.

5. **Tamperproof**: Does the system's protection system protect the trusted computing base programs?

   Scomp also uses protection rings and bracket model to protect the integrity of the rest of its trusted computing base. The Scomp trusted computing base runs in rings 0, 1, and 2. No untrusted code is supposed to run in these rings. It is possible for untrusted processes in ring 3 to invoke code in the trusted computing base. The number of interfaces and associated gates that were implement to protect the trusted computing base is unclear.

6. **Verifiable**: What is basis for the correctness of the system's trusted computing base?

   The Scomp system design and its implementation's correspondence to that design were verified with formal analysis tools.

7. **Verifiable**: Does the protection system enforce the system's security goals?

   Secrecy goals were enforced using a mandatory MLS policy for secrecy and discretionary bracket policies for integrity. Unlike prior systems, and most subsequent systems, the design of the system policies was also verified for correctness.

As a result, Scomp has mostly convincing answers to each of the questions above. While there are a few danger spots, such as the complexity of the interface to the trusted computing base, the possibility of modifications to the discretionary bracket policy, and the inherent incompleteness of system verification, Scomp and other security kernels are about as close to secure operating systems as possible. The challenges for security kernel systems are their performance, practical utility, and maintenance complexity.

6.3   GEMINI SECURE OPERATING SYSTEM

Another security kernel system that emerged in the 1980s was Gemini Corporation's Standard Operating Systems (GEMSOS) [3] [290]. The Gemini company was founded in 1981 with the aim of developing a family of high-assurance systems for multilevel secure environments. Gemini aimed to build its security kernel (GEMSOS) from scratch, but unlike Honeywell's Scomp system, implement it to run on the available commercial hardware, the Intel x86 architecture. The GEMSOS kernel and some systems based on the kernel were assured at TCSEC A1 level [109, 306], as Scomp was. Importantly, GEMSOS is still an active product, supported by the Aesec Corporation [5].

As the GEMSOS architecture has significant similarities to Scomp, we provide a higher-level description of its design, highlighting the similarities and key differences.

Architecture   The GEMSOS system architecture is shown in Figure 6.3. Like Scomp, GEMSOS consists of a security kernel and trusted software, but GEMSOS provides 8 protection rings, as does Multics. Since the x86 hardware only supports 4 rings, the security kernel runs in ring 0 and the trusted software in ring 1. The overall approach to security is derived from Multics as well, so secrecy protection is based on multilevel security labels and integrity is implemented by ring brackets.

GEMSOS does not define a kernel interface package, as Scomp does, but it does provide a library to make invoking the kernel gates easier, like the SKIP library. GEMSOS user-level processes (i.e., nonkernel code) access the security kernel using a Kernel Gate Library (KGL). This code runs outside the kernel, but because it may be used in user-level software that is part of trusted computing base, the KGL also is trusted code. This differs from the SKIP library which is not used in trusted software.

Security Kernel   The GEMSOS security kernel design consists of a layered set of kernel functionality, shown in Figure 6.4. This design was influenced by a number of ideas, including the layering approach of Reed [253] and the information hiding approach of Parnas [241]. The result is that each layer is dependent only on the layers below them, and the state of each layer is only accessible via well-defined interfaces.

The GEMSOS security kernel is constructed from a base of generic functions to provide typical kernel services including memory, I/O, and process management, in addition to reference monitoring. The lowest four layers provide generic functions for accessing the hardware, switching execution contexts, and interrupt handling among others. The Kernel Device Layer then provides

---

[3]GEMSOS may alternatively be called GSOS by some.

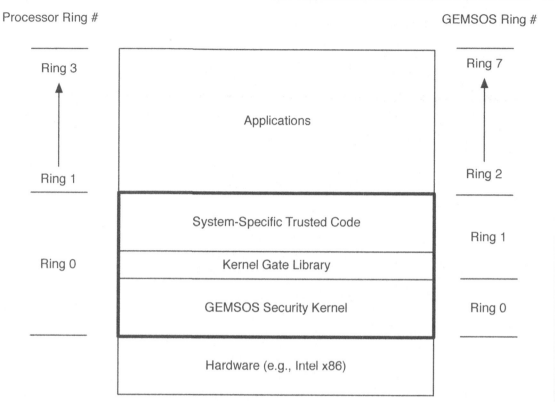

Figure 6.3: GEMSOS consists of a security kernel, gate library, and a layer of trusted software that is dependent on the deployed system. GEMSOS uses a software-based ring mechanism to simulate 8 protection rings.

other kernel layers with access to kernel-internal drivers. The Non-discretionary Access Control Layer implements the system reference monitor which enforces policies written in the Multics multilevel security model, see Chapter 3. The Secondary Storage Manager Layer provides the physical file system for GEMSOS user processes. Next comes the Internal Device Manager which provides the interface to device drivers. The Memory Manager Layer builds memory segments for kernel and user processes. The Upper Traffic Controller Layer provides support for multiprocessing using the concept of virtual processors. The top four layers, the Segment Manager Layer, the Upper Device Manager Layer, the Process Manager, and the Gate Layer all manage per-process resources: memory, I/O concurrency, processes, and system invocation, respectively.

The GEMSOS kernel architecture provides many of the services of ordinary kernels. But, the use of commercial hardware presented challenges to the designers. Because the x86 processor lacks the memory and device mediation of Scomp's Security Protection Module (SPM), device

Applications	
Gate Layer	↑
Process Manager (PM)	Process
Upper Device Manager (UDM)	Local
Segment Manager (SM)	↓
Upper Traffic Controller (UTC)	↑
Memory Manager (MM)	|
Inner Device Manager (IDM)	|
Secondary Storage Manager (SSM)	|
Non-Discretionary Security Manager (NDSM)	Kernel
Kernel Device Layer (KDL)	Global
Inner Traffic Controller (ITC)	|
Core Manager (CM)	|
Intersegment Linkage Layer (SG)	|
System Library (SL)	↓
Hardware	

Figure 6.4: GEMSOS Security Kernel Layers

drivers must be run in the GEMSOS kernel (e.g., in the Kernel Device Layer and Internal Device Manager). However, A1-level assurance requires verification of the correctness of all kernel (i.e., trusted computing base) code. Thus, as new devices and their drivers are introduced, this presents a management problem for the kernel. The availability of I/O MMUs [141, 8] would also enable the possibility of drivers outside the kernel.

The other major design similarity between GEMSOS and ordinary operating systems that differs from the Scomp is the presence of the file system in the kernel. In Scomp, the file system is implemented as part of the SKIP functional layer in ring 2. Recall that Scomp also included ring 2 software in the trusted computing base of the system. Later, researchers explored the design and implications of an untrusted file system on GEMSOS [146]. The GARNETS file system ran in a virtual machine outside the GEMSOS kernel, which results in an architecture similar to the Scomp approach. However, in the GARNET approach. the level of trust in the GARNET file system could be tangibly less than that of the kernel (i.e., it is not in the system TCB). The GARNET design required several workarounds to achieve the necessary functionality when this trust was removed, and may still require some trusted programs, albeit less trusted code than an entire file system.

GEMSOS defines 29 gates to access the security kernel, which is similar to the 38 gates provided by Scomp. The function offered by the gates are similar, although Scomp additionally provides function via the SKIP gates.

Trusted Software   GEMSOS also provides a set of trusted software running outside the security kernel. The functions of the GEMSOS trusted software are similar to those provided by Scomp trusted software. For example, there are trusted software services for system administration.

Applications   GEMSOS differs from Scomp in the number and variety of applications in which it was deployed. GEMSOS was commonly applied as a platform for securely connecting networked high security systems and isolating them from low security systems in the same network [242]. Also, GEMSOS was applied as a guard for an office software suite.

The most significant applications of GEMSOS were for network control. First, GEMSOS was the basis for the multilevel secure system called BLACKER [330]. BLACKER provided an A1-assured component for key distribution and secure communication to protect high secrecy data in transit between high secrecy networks. BLACKER consists of a set of encryption devices that enable the isolation of a high secrecy network from the rest of the Internet. Originally, different processors were used to handle the ciphertext (the *black* side) and the plaintext (the *red* side). Of course, it is important that ciphertext not leak the contents of plaintext, but if the ciphertext is created on a system with a Trojan horse, this cannot be guaranteed.

Second, GEMSOS was later applied to the general notion of a Trusted Network Processor (TNP) [306]. A TNP hosts one or more applications that require multilevel security enforcement. The applications themselves run on virtual processors in a multiprocessor system (potentially), but they are ignorant of multilevel security. The labeling of virtual processors and data and the enforcement of multilevel security are performed by the TNP component that mediates all communication on the multiprocessor bus.

A POSIX interface was developed for the GEMSOS system as well, although historically GEMSOS was applied to dedicated or embedded applications.

6.4   SUMMARY

The secure operating systems that followed the Multics system focused on the key limitations of Multics: performance and verifiability. The idea of a *security kernel*, a small kernel with minimal code in its trusted computing base, addressed both of these problems. First, the design of security kernel was customized to address performance bottlenecks, even by adding security function in hardware, such as Scomp's Security Protection Module. Second, the small size of the security kernels also motivated the development of system assurance methodologies to verify that these systems correctly implemented a secure operating system (see Definition 2.5). Multics was too large and complex to be verified, but both Scomp and GEMSOS were of manageable complexity such that the verification tools of the day, plus some manual effort, were sufficient to justify the correctness of these implementations. Such efforts led to the development of the assurance methodologies that we use today, see Chapter 12.

Security kernels are general purpose systems. The design of Scomp provides similar system function to UNIX systems of the day, and GEMSOS is a full kernel. Nonetheless, security kernels became niche systems. The performance, flexibility, and applications in UNIX systems and, later, Windows systems limited the market of security kernels to specialized, high security applications, such as guards. Further, the need to balance assurance and function became difficult for these security kernels. Maintaining the assurance of the kernel given the vast number of drivers that are developed, including some which are quite buggy, is very difficult. Scomp did not depend on drivers in the kernel, but it did depend on hardware features that were not available on common processors.

The need identified by security kernels has continued to exist. Scomp was succeeded by the XTS-300 and XTS-400 systems, now distributed by BAE [22]. GEMSOS is still available today from Aesec [5]. There have been other security kernel systems, including Boeing's secure LAN [298], Secure Computing Corp.'s LOCK system [293], and KSOS [198]. Also, the separation kernel systems (see Chapter 11) aim for a minimal, assured trusted computing base for deploying applications, and this architecture is also used frequently. As a result, it appears that the development of a minimal platform necessary to deploy the desired software is still the preferred option for security practitioners, although as we will see in next chapters, secure system alternatives that already support the desired applications becoming more popular and more secure.

CHAPTER 7

Securing Commercial Operating Systems

Since the discovery of the reference monitor concept during the development of Multics, there have been many projects to retrofit existing commercial operating systems with a true reference monitor implementation. Successful, commercial operating systems can have a large customer base and a variety of popular applications. As a result, those customers with strong secrecy and integrity requirements (e.g., US Government) often encourage the construction of secure versions of existing commercial operating systems. Many such systems have been retrofitted over the years.

In this chapter, we explore some of the commercial systems that have been retrofitted with reference monitors. The aim is not for completeness, as there are far too many systems, but we want to capture the distinct movements in creating a secure operating system from an existing commercial system.

Converting an existing code base to one that implements a reference monitor is a challenging task. In order to be a secure operating system, the resulting code base must achieve the three reference monitor guarantees, but this is difficult because much of the code was not developed with these guarantees in mind. This contrasts markedly with the *security kernel* approach in Chapter 6 where the system design considers mediation, tamperproofing, and verification from the outset.

After outlining the tasks involved in retrofitting a commercial operating system with a reference monitor, we examine a variety of different retrofitted systems. We group these systems by a trend motivating their construction. We examine the resultant system architectures in detail for two systems: Solaris Trusted Extensions in Chapter 8 and the Linux operating system in Chapter 9.

7.1 RETROFITTING SECURITY INTO A COMMERCIAL OS

To retrofit a commercial operating system into a secure operating system, the resultant operating system must be modified to implement a secure operating system that implements the reference monitor concept, see Definitions 2.5 and 2.6. The reference monitor concept requires guarantees in complete mediation, tamperproofing, and verifiability. There are challenges in each of these areas.

Complete mediation requires that all the security-sensitive operations in the operating system be identified, so they can be authorized. Identifying security-sensitive operations in a complex, production system is a nontrivial process. Such systems have a large number of security-sensitive operations covering a variety of object types, and many are not clearly identified. As we will see, a significant number of security-sensitive operations are embedded deep inside the kernel code. For example, in order to authorize an open system calls, several authorizations may be necessary for directories, links, and finally the target file (i.e., `inode`) itself. In addition to files, there are many such objects in modern operating systems, including various types of sockets, shared memory, semaphores, interprocess communication, etc. The identification of covert channels (see Chapter 5) is even more complex, so it is typically not part of retrofitting process for commercial operating systems. As a result, complete mediation of all channels is not ensured in the retrofitted operating systems we detail.

Tamperproofing the reference monitor would seem to be the easiest task in retrofitting an existing system, but this also has proven to be difficult. The obvious approach is to include the reference monitor itself in the kernel, so that it can enjoy the same tamper-protection that the kernel has (e.g., runs in ring 0).

There are two issues that make guaranteeing tamper-protection difficult. First, commercial operating systems often provide a variety of ways to update the kernel. Consider that UNIX kernels have a device file that can be used to access physical memory directly `/dev/kmem`. Thus, processes running outside of the kernel may be able to tamper with the kernel memory, even though they run in a less-privileged ring. Modern kernels include a variety of other interfaces to read and write kernel memory, such as `/proc`, `Sysfs` file systems, and `netlink` sockets. Of course, such interfaces are only accessible to `root` processes, but there are many processes in a UNIX system that run as `root`. Should any one get compromised, then the kernel may be tampered. In effect, every `root` process must be part of a UNIX system's trusted computing base to ensure tamper-protection.

But the biggest challenge for retrofitting an operating system is providing verification that the resultant reference monitor implementation enforces the required security goals. We must verify that mediation is implemented correctly, that the policy enforces the expected security goal, that the reference monitor implementation is correct, and that the rest of the trusted computing base will behave correctly. Verifying that the mediation is done correctly aims to address the problems discussed above. Typically, the mediation interface is designed manually. While tools have been developed that find bugs in mediation interfaces [149, 351], proving the correctness of a reference monitor interface in an operating system is intractable in general because they are written in nontype safe languages, such as C and various assembly languages.

Policy verification can also be complex as there are a large number of distinct authorization queries in a commercial operating system, and there are a large number of distinct processes. Some retrofitted commercial operating systems use a multilevel security (MLS) model, such as Bell-LaPadula [23], but many use access matrix mandatory access control (MAC) models, such as Type Enforcement [33]. The latter models are more flexible, but they also result in more complex policies. A Bell-LaPadula policy is fixed size, but an access matrix policy tends to grow with the number of distinct system programs. Such models present a difficult challenge in verifying that each system is enforcing the desired security goals.

Finally, the implementation of a commercial operating system and the remaining trusted computing base is too complex to verify whether the overall system protects the reference monitor. Commercial operating systems are large, there are often several developers of the trusted computing base software, and the approaches used to build the software are not documented. The best that we can hope for is that some model of the software can be constructed after the fact. As described in Chapter 6, the verification of Scomp's correctness required an evaluation that the design model enforced system security goals and that the source correctly implemented the design. Many believe that it is not possible to build a sufficiently precise design of a commercial system and a mapping between this design and the system's source code necessary to enable such verification. Clearly, current technologies would not support such a verification.

7.2   HISTORY OF RETROFITTING COMMERCIAL OS'S

In this section, we examine the evolution of retrofitting security into commercial operating systems. We organize this section by identifiable eras in the construction of secure operating systems. As the lessons from the Multics project were being disseminated, many companies examined ways to retrofit Multics-style security into their existing commercial operating systems during the *commercial era*. The invention of the microkernel systems led to several attempts to retrofit security in the smaller, microkernel architectures, which resembled security kernels (see Chapter 6), during the *microkernel era*. Gradually, the focus returned to UNIX systems, which had become the de facto server operating system (although there were many distinct UNIX systems maintained by competing entities by then). Some of the novel ideas of the commercial and microkernel era were transferred to UNIX-style systems in this most recent era, the *UNIX era*.

Each of the eras focused on particular themes. The commercial era work focused on either emulation of commercial systems on security kernels or retrofitting by adding orthogonal features to existing code bases. The result of this era was systems that enforce multilevel secrecy policies in UNIX. The microkernel era focused on adding security via independent server processes, but as the work proceeded, more invasive modifications, lower in the software stack were deemed necessary. Also, innovative security models emerged that aimed to address both secrecy and integrity comprehensively. The UNIX era composed the mature solutions of the first two eras with a renewed focus on system integrity. Both Solaris Trusted Extensions (see Chapter 8) which resulted from the commercial era and the SELinux (see Chapter 9) which resulted from the UNIX era have adopted many similar solutions, although there are significant differences and some challenges remain open to future research.

7.3   COMMERCIAL ERA

In the late 1970s and early 1980s, it became clear that Multics provided some fundamental security features, see Chapter 3, but it was too large and slow to be effective. A variety of competing vendors saw improvements in security as a potential advantage for their systems. A goal became to capture Multics security features in their commercial systems. The chief question was how to marry the security enforcement of Multics with the application interface of these commercial system.

Emulated Systems: Data Secure UNIX and KSOS   Some projects focused on the construction of a security kernel, see Chapter 6, that ran an emulator for the UNIX API. UCLA Data Secure UNIX [248] and KSOS [198, 97], fall into this category. In both cases, the performance of the emulated systems was poor, so later security kernels, such as Scomp, dropped the idea of an emulator.

These systems did not really integrate security into the existing operating system, but rather tried to slide a secure environment under the existing system. However, in addition to problems in performance, insecure features of the UNIX interface, such as permitting the passing of file descriptors on process creation regardless of the relationship between the processes, also presented security problems that could cause incompatibilities with the security kernel.

KVM/370   The KVM/370 system adds a layer between the virtual machine monitor (VMM) and individual virtual machine (VM) to mediate inter-VM communication. The design of this new layer is a retrofit of the existing VM/370 code base with multilevel security features. The retrofit resulted in performance overhead of about 25% of a typical VM/370 virtual machine. This was partly due to the additional layer between the virtual machines and the VMM and partly due to required reuse of VM/370 code in the new KVM/370 system, which introduced extra effort in indirection. Virtual machine-based secure operating environments are discussed in Chapter 11.

VAX/VMS   DEC and Sandia Labs retrofitted VAX/VMS with multilevel security enforcement [180]. In addition, improvements in auditing were developed and a number of security vulnerabilities were fixed. Because of the retrofit of the existing code base, the VAX/VMS system aimed only for modest assurance levels, B1 or B2 in the Orange Book [304]. This work was a prototype and performance impact was not discussed.

Secure Xenix   Somewhat later than the work above, IBM retrofitted Microsoft's Xenix with access control and auditing features [111]. This work was influenced the UNIX retrofit of Kramer [173], but aimed to provide a comprehensive and effective implementation of Multics security features [280] (see Chapter 3) in Xenix. Two key issues among several addressed by the Secure Xenix work were compatibility and trusted path. First, the Secure Xenix system included both the retrofitting of a variety of UNIX services with security-aware function, so that UNIX applications could be run without modification. Further, compatibility mechanisms, such as hidden subdirectories, were invented to enable multiple processes at different security levels to "shared" directories without introducing information leakage. This mechanism is the basic idea behind *polyinstantiated file systems* discussed in Section 8.2 and used in several systems now. Second, Secure Xenix also introduced the notion of a *trusted path*. A *trusted path* is a mechanism to communicate directly with the system's trusted computing base. A trusted path is often implemented via a "secure attention sequence" that can only be caught by the trusted computing base (e.g., Control-Alt-Delete). Thus, the user can be certain that she is communicating with trusted code. Secure Xenix was successfully evaluated at the US government B2 rating based on the Orange Book [304]. Secure Xenix was later renamed Trusted Xenix, when its development was shifted to Trusted Information Systems.

By 1990, a variety of UNIX variants had been extended with security mechanisms, particularly those aiming at MLS enforcement [339]. One of these systems, SunOS MLS was introduced in 1989, but it ultimately established itself as the market leader in MLS systems. It has continued to evolve over the last twenty years, so we present an overview of the current version, called Trusted Solaris Extensions, in Chapter 8.

7.4   MICROKERNEL ERA

In the 1980s, microkernel systems emerged. Microkernel systems were similar to security kernels in that they aimed for minimal functionality in the kernel, but microkernel systems focused on providing system abstractions for building complete systems more easily and more efficiently, rather than more securely. The hope was that microkernel systems would be more effective at running UNIX systems (i.e., perform better), while preserving the economy of size and potential for verification offered by the security kernels.

The emergence on the Mach microkernel [348, 116] in the 1980s was the source of interest in microkernel systems. Mach aimed to be a minimal kernel while provided abstractions to enable complete operating system construction, including mechanisms for message passing between components and multi-threaded process support. However, fundamental operating systems services, such as memory managers, file systems, and network servers, are not implemented in the microkernel, so user-level servers must be designed to implement such functions. Thus, the system's trusted computing base consists of the microkernel and those user-levels servers that must be trusted. While these microkernel architectures appear similar to the security kernel architectures of Chapter 6, typically microkernel systems were built to improve nonsecurity dimensions of operating systems, such as ease of development, flexibility, and even performance.

Several projects that used Mach as the base for a secure operating system were inspired by that architecture. These included Trusted Mach (TMach) [36, 35, 196], Distributed Trusted Mach (DTMach) [282, 96], the Distributed Trusted Operating System (DTOS) [313, 213], and the Flask system [191, 295]. TMach was built by Trusted Information Systems (TIS) and implemented multilevel security (MLS) servers for files, memory, etc. that would provide function for single-level operating system personalities, such as UNIX or Windows. Thus, TMach provides trusted services for MLS computing where each instance of a traditional operating system runs as a single-level system. DTMach was built by Secure Computing Corporation (SCC) and the National Security Agency (NSA). DTMach supports a hybrid access control model that uses both MLS labels for secrecy control and Type Enforcement (TE) [33] labels for integrity. TE is an access matrix-based, mandatory access control policy with a fixed set of subject and object labels, called *types*, and the policy defines which subject types may perform which operations on which object types. TE was first applied in the LOCK system [293]. When a new file or process is defined, it is labeled from the type set and inherits the policy defined by the TE matrix.

The DTMach architecture is similar to TMach, but it also includes additional servers for networking DTMach systems and providing general security policy server support. DTOS was the SCC/NSA/University of Utah followup to the DTMach system. The DTMach project found security limitations in the Mach microkernel mechanisms that the DTOS project aimed to fix. The Mach architecture was found to have significant performance issues, so the DTOS architecture was migrated to another microkernel system for the Flask project. We explore some of the issues in DTMach, DTOS, and Flask projects below.

DTMach   DTMach extended Mach with a separate security server, a reference monitor outside the kernel that responds to authorization queries. As file, network, and interprocess communication (IPC) are invoked by sending messages to Mach *ports*, DTMach authorization queries are invoked on port access. For example, when a process opens a file, it sends a message to a port of the file server hosting that file. The security server is invoked to ensure that the process has the necessary permissions to access the file.

The DTMach security server represents permissions in two forms, MLS permissions and TE permissions. MLS permissions enforce secrecy using the traditional Bell-LaPadula model [23]. TE permissions were used to protect the integrity of the system. TE was used in DTMach to define limited mandatory domains for users and particular system services. TE policies in DTMach limit code installation and modification to administrators only, limit the code that can be executed by system subjects, prevent servers from having unnecessary rights to system objects, ensure that only authorized downgraders could relabel certain data, etc.

However, Mach ports suffered from some limitations that prevented correct enforcement of TE policies. For example, a `send` right on a Mach port implies that a process with that right can send arbitrary messages to the port, but we may want to limit the set of messages that untrusted processes can send to ports served by trusted processes. Consider when an untrusted process asks for a file to be mapped into its address space. In this case, the process must have a `send` permission to the memory pager to ask for the file to be mapped. However, this right permits the untrusted process to send any message to the pager, increasing the complexity of pager. DTMach defines more nuanced `send` rights to only allow file mapping requests. There are other specific cases where the meaning of a `send` permission to a port must be limited, and these are all handled by extending the port authorization mechanism.

DTOS   The changes to Mach and its server to address port control resulted in several ad hoc changes to the Mach microkernel. In the DTOS project, the aim was to construct a true reference monitor in the Mach microkernel [283, 213]. To address the problems caused by `send` port permissions above, DTOS defined a richer set of operations for operating on ports. The DTOS Mach microkernel managed the labeling of subjects and kernel objects and provided access control over each kernel operation by querying the security server itself. This resulted in complete mediation of kernel operations with the richness necessary to limit access to trusted servers in a tamperproof Mach microkernel. The DTOS project also focuses on verifiability through assurance of the microkernel and its trusted computing base.

Fluke/Flask   So-called second generation microkernels made significant improvements in the IPC performance [184, 185, 83] making Mach obsolete, so the DTOS architecture was ported to a second generation microkernel called Fluke from the University of Utah [98]. The resulting security architecture, called Flask, retained many of the elements of the DTOS architecture, in particular the microkernel reference monitor, trusted servers (called *object managers*), and a separate security server [295]. At this time, the focus shifted from MLS to the TE mechanism, as the latter is more general and integrity protection became the focus. As with all the second generation microkernels, Fluke did not attract a large user community, and another UNIX-based system was emerging in popularity, Linux, see Chapter 9.

7.5   UNIX ERA

By the early 1990s, a variety of different approaches to retrofitting UNIX systems has been explored [339]. In the process of constructing these systems, a variety of technical challenges were discovered and solutions were proposed. This ultimately resulted in successful deployment of UNIX systems that supported MLS policies. However, the research community continued to explore comprehensive UNIX retrofitting that would address both integrity and security in concert. While the UNIX MLS systems did not ignore integrity, integrity was addressed more implicitly, so it was left to the administrators to ensure that the inputs to their high secrecy systems were also high integrity. We examine two systems in detail here, IX and DTE, which retrofit UNIX with integrity and secrecy mechanisms.

7.5.1   IX
AT&T Research built an experimental UNIX prototype that enforces multilevel (MLS) secrecy and integrity, called IX [200]. IX provides a reference monitor over file access that implements a mandatory access control policy that provides secrecy and integrity protections. Care is also taken in the definition of the trusted computing base to prevent tampering. Verification of security guarantees is not a focus in IX. This is partly due to the complexity of verifying correctness on an existing kernel, and partly due to the desire for more flexible labeling.

   The IX mandatory access control policy enforces information flow secrecy. Processes have labels in a secrecy lattice that ensure that they may only read data at their secrecy level or lower. Unlike traditional MLS systems, such as Multics [23, 280], IX uses dynamic labeling to provide more flexible information flow control, however. The label of a file may change if it receives information from a process at a higher secrecy level in the lattice. For example, a file may start with a low secrecy label, but its label is changed to high secrecy when a high secrecy process updates it.

IX also includes a transition state that enables relabeling of processes and objects. Process labels change based on the secrecy labels of the files that they have read. However, a *ceiling* is defined on process labels to limit the level that a process may reach. Ceilings may also be associated with file systems to limit the secrecy of data written to that file system.

IX also supports a separate integrity lattice with dynamic labels as well. IX uses the LO-MAC [27, 101] semantics for integrity labels. The label of an entity (process or file) is equal to the lowest lattice label of an input to that entity. IX uses label *floors* to limit the degradation of integrity that is possible for particular processes or files. A process may not read a file whose integrity is lower than its floor which ensures that the integrity of a process is at least at the label of its floor.

The use of dynamic labeling requires authorization on each data transfer. For example, if a file is opened for writing when the process is at high integrity, it may no longer be written after the process's integrity has decreased. Thus, authorization must be performed on all reads and writes, not just at the time the file is opened. Such semantics precludes the use of memory-mapped files, because file accesses are implemented by memory operations, rather than system calls. The risk is that a process could map a high integrity file into its address space, load potentially malicious code, and write to the file via the mapped memory without mediation by the reference monitor. Thus, memory-mapped files must be prohibited in IX.

IX provides mechanisms for establishing trusted paths (called *private paths* in IX) between trusted processes. IX defines a *pex*, a process exclusive access to a file or pipe, that prevents interference from other processes. Further, a pex provides labels of the processes on each end of the pex, such that an *assured pipeline* [33] of processes can be constructed. Figure 7.1 demonstrates an assured pipeline. *Process 1* is the only one that can read the input data (labeled *0th* and generate the output data (labeled *1st*). *Process 2* then reads the *1st* data and then outputs the next data *2nd* that is used by *Process 3*. Since an assured pipeline can be used to ensure that a computation spanning multiple processes is high integrity, each of its stages must be performed only by trusted code.

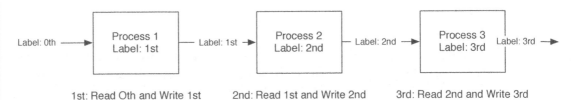

Figure 7.1: An *assured pipeline* chains together processes to perform a secure computation where each process can only communicate with its predecessor and successor. *Process 1* is the only one that can read the input data labeled *0th*, and outputs data of label *1st* that can only be read by *Process 2*.

7.5.2   DOMAIN AND TYPE ENFORCEMENT

Trusted Information Systems (TIS) retrofitted UNIX with a reference monitor that implements an extension of the Type Enforcement (TE) [33] policy, called Domain and Type Enforcement

(DTE) [15, 16]. UNIX has many trusted (i.e., root) processes that had been found to be vulnerable (and often still continue to be vulnerable) to malicious network inputs. The DTE approach aims to confine UNIX processes to protect the trusted computing base from other root processes.

Strictly speaking, DTE UNIX runs as a server on a TMach system [35, 15]. However, we consider DTE to be a retrofit of UNIX because the reference monitor is added to the UNIX server (OSF/1), not to TMach.

DTE Policy Model   DTE extends the TE model by distinguishing subject types from object types and adding transition states. In classical TE [33], there is only one set of types covering all subjects and objects. In a DTE policy, types are assigned to objects, and *domains* are assigned to subjects (i.e., processes). A domain is a tuple consisting of three parts: (1) access rights to object types; (2) access rights to subjects in other domains (e.g., signals); and (3) an entry point program, a file that when executed triggers the domain. DTE domains describe how a process accesses files, signals processes in other domains, and creates processes in another domain.

First, DTE UNIX defines limited protection domains for root processes [323]. Instead of complete access, a least privilege policy is defined for such processes. Thus, should a network-facing daemon be compromised, the extent of the damage that the attacker could cause may still be limited. For example, it may not be possible for a compromised root process to install a rootkit under the confinement of the TE policy. Also, the definition of domain transitions (see below) limits which domains may be invoked by any process, also limiting possible malicious actions should a process be compromised.

Second, *signals* are a mechanism for the operating system or other processes to notify a process. A signal interrupts the target process and forces it to handle the signal immediately. Signals may be used for a variety of purposes, such as terminating or resuming a process, but the process's signal handler defines the effect. If an untrusted process can submit a signal to other processes, then it can cause unauthorized execution resulting in termination or incorrect behavior.

The third element in domain specification defines the transition state. For each domain, DTE enables us to control transitions using the third element of a domain specification. In DTE, we limit domain transitions to the execution of a specific file corresponding to that domain and limit the domains that can cause that transition. In UNIX, the conditions under which a process starts also depends on environment variables, input arguments, file descriptors not set to close-on-exec, etc., so controlling the file used for a domain transition is only a start. Compare this to a ring transition in Multics which also uses gatekeepers to verify arguments, see Chapter 3.

Labeled Networking   A key innovation that appeared around this time was *labeled networking*. The idea is that each machine labels its network packets so that the receiver may authorize delivery to its processes. For example, two machines may have both secret and confidential processes. When a secret process sends a packet, the sending operating system adds the label to the packet header. When the packet is received, the receiving operating system extracts the packet's label, and authorizes delivery to a receiver's socket based on its label. Thus, if the `secret` packet is targeted for a socket created by a `confidential` process, the receiving system can deny authorization. Thus, authorization can span processes on multiple machines.

Labeled networking was enabled by the addition of IP security options to the IP header and protocol [153], later revised in IETF RFC 1108 [165]. IP security options enabled transmission of the sensitivity level of the packet. Implementations that used the IP security options header were developed. Ones based on RFC 1108 directly are referred to as revised IP security options (RIPSO) system. Since the information is simply stored in the IP header, external mechanisms were still needed to protect the header in transit over an untrusted network (e.g., IPsec [168, 166, 167]).

A later revision to the approach, called the commercial IP security option (CIPSO), was proposed and implementations were built to support it, although it was never standardized (there is a draft [53]). CIPSO has generally superseded RIPSO. It consists of a domain of interpretation, defining the meaning of the labels, and a specification for labels (tags) that can include levels and categories. Both secrecy and integrity levels are supported. Implementations often support both RIPSO and CIPSO. Solaris Trusted Extensions (see Chapter 8) supports RIPSO, CIPSO, and its own variant called Trusted Systems Information Exchange for Restricted Environments (TSIX) [301].

DTE also includes a form of labeled networking [15]. In this case, DTE includes the labels of the data being written (type) and the process doing the writing (domain) inside the packet. IP security options are also used to encapsulate the DTE labels. The labeling of both process and data domains means that the system must be able to reliably label the data being sent. For example, NFS servers were modified to use DTE networking, and they are responsible for labeling the file data that they deliver to clients. The range of labels that may be specified by a particular NFS server may be limited by DTE, so there is not a need for fully trusted servers.

7.5.3   RECENT UNIX SYSTEMS

A variety of UNIX systems now include a significant set of security features. We focus on two such UNIX systems in this book, Solaris Trusted Extensions in Chapter 8 and SELinux in Chapter 9, but there are several others of note. In particular, security has been a major focus of several of the BSD variants. The BSD systems derived from the Berkeley Software Distribution, a derivative of UNIX developed at the University of California, Berkeley [201]. We briefly examine them here, but recommend a detailed examination of their security features, as well.

Within the FreeBSD [103] community, the Trusted BSD project [327, 328, 312] aims to implement trusted operating system extensions for FreeBSD. TrustedBSD includes services for mandatory access control, auditing, and authentication, necessary to implement a mandatory protection system (see Definition 2.4 in Chapter 2). TrustedBSD includes a mandatory access control framework that implements a reference monitor interface, analogous to the Linux Security Modules framework [342] for Linux (see Chapter 9). The TrustedBSD framework enables the development of reference monitor modules that can enforce mandatory access control policies. One such module is SEBSD, a version of the SELinux module designed for BSD. Another FreeBSD security project is FreeBSD Jails [156], which implement a lightweight virtualization similar to Solaris Containers (see Chapter 8) where processes running in a jail are isolated from processes outside that jail.

The security focus for OpenBSD [235] is on correct coding and configuration of systems to minimize its attack surface. OpenBSD does not enforce mandatory access control, but instead focuses on the correctness of its trusted programs and limiting the amount of code in trusted programs. Rigorous code reviews are required for all trusted programs to reduce the possibility of vulnerabilities. Privilege separation is often employed to re-engineer trusted programs to remove code that does not require root privileges to execute, such as for OpenSSH [251]. This approach reduces the trusted computing base of the system, by ensuring that less code has the privileges necessary to compromise the system. In addition to code review and privilege separation, other system hardening techniques, such as buffer overflow protection and least privilege configurations, are employed to prevent system compromise. As a mandatory protection system requires that a tamperproof trusted computing base, such attention to the trusted computing base programs is necessary for a secure system in general. Other systems can, and have, leveraged the re-engineered programs developed for OpenBSD.

NetBSD [224] contains many of the security features of modern UNIX systems to prevent buffer overflows, but it additionally provides in-kernel authentication and verification of file execution. In UNIX, user authentication is traditionally performed by trusted programs running outside the kernel. These programs are vulnerable to compromised root programs (e.g., network facing daemons), so the system's security may depend on programs that cannot be protected from tampering. As NetBSD's Kauth framework is deployed inside the kernel, it is not susceptible to compromised user-space processes, so trust in authentication is improved. NetBSD also defines a Veriexec mechanism which can be used to verify the integrity of a file prior to its use. Veriexec ensures that only files whose contents correspond to an authorized hash may be accessed. It defines different modes of permissible access for a file: (1) DIRECT for executables; (2) INDIRECT for interpreters run indirectly (e.g., via #!/bin/sh); and (3) FILE for data files that may not be executed. The NetBSD kernel checks the integrity (i.e., hash) of the file before its is accessed in the specified manner to detect unauthorized modification.

7.6   SUMMARY

Adding security features to an existing operating system, with its existing customer base and applications, has been a popular approach for building secure systems. Unfortunately, retrofitting security into existing, insecure systems leads to a variety of issues. Many programs are designed and configured such that they will not work in the more restrictive environment of a secure system. The operating systems themselves have complex interfaces that may be difficult to mediate.

In this chapter, we surveyed a variety of systems where security is retrofitted. We describe the security features that are added to these systems, the challenges in ensuring that the reference monitor concept is achieved, and the decisions that were taken to address these challenges. In general, these efforts show that it is practical to add a reference monitor interface to an existing system, but that it difficult to ensure the reference monitor guarantees are actually achieved. The complexity and dynamics of these commercial systems prevent security professionals from developing models necessary to verifying mediation, tamperproofing, or correctness. We examine these challenges in detail for Solaris Trusted Extensions in Chapter 8 and for Linux in Chapter 9.

CHAPTER 8

Case Study: Solaris Trusted Extensions

Glenn Faden and Christoph Schuba, Sun Microsystems, Inc.

Solaris (TM) Trusted Extensions is a feature of the Sun Microsystems's Solaris operating system that enforces multilevel security (MLS) policies [23]. It is the latest in a series of MLS workstation and server operating systems that have been under development at Sun since 1988. The first version, SunOS MLS 1.0, which appeared in 1990, was based on the SunView window system. It was designed to meet the TCSEC B1 level [304], see Chapter 12. However, it was replaced in 1992 by SunOS CMW, which was designed to meet the Compartmented Mode Workstation Requirements, CMWREQS [25, 340]. SunOS CMW was based on OpenWindows and X11/NeWS. It supported both sensitivity labels for Mandatory Access Control, and floating information labels for human consumption. It was first certified using the ITSEC Scheme at the E3/FB1 level in 1992.

Trusted Solaris 2.5 through Trusted Solaris 8 were based on the Common Desktop Environment (CDE) and X11 [227]. Trusted Solaris 2.5.1 was also certified using the ITSEC scheme at the E3/FB1 level in 1996. Trusted Solaris 8 was evaluated using the Common Criteria scheme in 2000, with an assurance level of EAL4+. It was certified to meet the Controlled Access (CAPP) [230], Role-Based Access (RBACPP) [256], and Label Security Protection (LSPP) [231] Profiles. The RBAC features of Trusted Solaris were incorporated into the standard Solaris OS at that time. Assurance is detailed in Chapter 12, but in general, the assurance validates the correct low-level design for enforcing MLS requirements. Based on this assurance, Trusted Solaris has a dominant share in the U.S. Department of Defense and intelligence communities.

In 2001 Sun began work to unify its two Solaris versions, which was completed in 2006, with the release of Solaris 10, update 3, which included the Trusted Extensions [1]. Also at that time, Sun contributed the source code for the kernel and window system to the OpenSolaris community. In addition to removing the need for separate kernels, the integration also made it possible to support MLS on x86, x64, and SPARC platforms. Trusted Extensions includes an MLS version of the GNOME desktop. The combined Solaris system with Trusted Extensions received Common Criteria certification at the EAL4+ assurance level in June 2008, using the same three protection profiles.

The authors of this chapter would like to thank their colleagues in the Sun Solaris Security Organization, especially Casper Dik, Gary Winiger, Darren Moffat, and Glenn Brunette, for their contributions and reviews.

This new approach enables the Solaris operating system to support both traditional Discretionary Access Control (DAC) policies based on ownership, as well as label-based, Multilevel Security (MLS) policies. The MLS label-based policies for file systems and networks have been implemented throughout the standard Solaris 10 kernel, its services and utilities. Unless the Trusted Extensions layer is enabled, all labels are equal, so the kernel does not have any MLS requirements to enforce.

The Trusted Extensions systems provide a reference monitor implementation for Solaris that enforces an MLS policy. The reference monitor extends the Solaris (and traditional UNIX) enforcement by providing complete mediation and extending file enforcement to network, printing, and devices. Further, Trusted Extensions provides extensive support for labeling objects in the first place. Trusted Extensions does not need to enable transition of process or resource labels, a mechanism commonly used in Domain Type Enforcement (DTE). Tamperproofing is improved by reducing the rights on root processes, using limited domains similar to those in DTE. Finally, verification of correctness is limited, as for all retrofitted UNIX systems, by the amount of code reused from insecure systems. However, the focus of the security policy is primarily on secrecy, so the correctness of the secrecy policy can be verified, but understanding the integrity of the system data is an ad hoc process.

The trusted computing base of Trusted Extensions included the kernel and a variety of administrative applications. Importantly, a variety of administrative applications also have to be modified to be MLS-aware, so that they can assist the operating system in the enforcement of MLS requirements. For example, authentication services must be capable of determining MLS labels for users as they login.

8.1 TRUSTED EXTENSIONS ACCESS CONTROL

The Trusted Extensions access control model supports secrecy protection via MLS, process confinement in a manner similar to DTE, and ad hoc privileges to work around limitations of the first two policies. First, both sensitivity levels and categories are used to describe the possible information flows in a system. Second, Trusted Extensions adds roles for limiting the rights of processes that traditionally ran as root, like domains in DTE. As a result, root is only used at installation time, so no processes run with full privilege. Third, discrete rights exceptional to the above two policies may be granted to an application using Solaris privileges. There are at least 68 different kinds of discrete privileges that may be granted.

The default mandatory policy of Trusted Extensions is a mandatory, multilevel security (MLS) policy that is equivalent to that of the Bell-LaPadula Model [23] of the Lattice, the Simple Security Property, and the ⋆-Property (Star Property), with restricted write up. The default mandatory policy is also equivalent to the Goguen and Mesegeur model [113] of Non-Interference.

Labels consist of hierarchical components called classifications (or levels) and a nonhierarchical components called compartments (or categories). The mapping of names to classifications and compartments is specified in a database which is private to the Trusted Path. The internal structure of labels is deliberately opaque to users and applications and might change in a future release. At least 256 classifications and 256 compartment bits are supported.

When two labels are compared, the first label can be greater than, less than, equal to, or disjoint from the second label. Classifications are compared as integers, and compartments are compared as bit masks. Labels are disjoint when each contains at least one compartment bit which is not present in the other.

A label range can be specified by an upper bound (called a clearance), and a lower bound. Administrative roles can use the Trusted Path to assign label ranges to users, network attributes, workstations, and devices.

For this MLS policy, two labels are always defined: `admin_low` and `admin_high`. The site's security administrator defines all other labels. The label `admin_low` is associated with all normal user readable (viewable) Trusted Extensions objects while the label `admin_high` is associated with all other Trusted Extensions objects. Only administrative users have MLS read (view) access to `admin_high` objects and only administrative users have MLS write (modify) access to `admin_low` objects or `admin_high` objects.

Users interact with labels as strings. Graphical user interfaces and command line interfaces present these strings. Human-readable labels are classified at the label that they represent. Thus, the string for a label A is only readable (viewable, translatable to or from human readable to an opaque label data type) by a subject whose label allows read (view) access to that label.

In order to store labels in publicly accessible (i.e., `admin_low`) name service databases, an unclassified internal text form is used. This textual form is not intended to be used in any interfaces other than those that are provided with the Trusted Extensions software release that created this textual form of the label.

When the label-based MLS policies are enabled, all data flows are restricted based on a comparison of the labels associated with the subjects requesting access and the objects containing the data. Like other multilevel operating systems, Trusted Extensions meets the requirements of the Common Criteria's Labeled Security Protection Profile (LSPP) [231] and the Role-Based Access Protection Profile (RBAC) [256]. The Trusted Extensions implementation focuses on maintaining compatibility with existing UNIX systems, maintaining the high performance of UNIX systems, and simplifying administrative tasks.

8.2   SOLARIS COMPATIBILITY

Compatibility with thousands of Solaris applications is achieved by building on existing Solaris features and using existing industry standards. No new protocols are required nor new file system attributes. Applications do not need to be modified nor profiled to bring them into conformance with the MLS policy. Instead, the entire application environment is virtualized for each label through the use of Solaris Containers (zones) [175], Solaris primary OS-level virtualization technology. The Solaris Containers facility provides an isolated environment for running applications. Processes running in a zone are prevented from monitoring or interfering with other activity in the system.

Access to other processes, network interfaces, file systems, devices, and inter-process communication facilities are restricted to prevent interaction between processes in different zones. At the same time, each zone has access to its own network stack and name space, enabling per-zone network security enforcement, such as firewalling or IPsec.

The approach to map labels with zones is referred to as *polyinstantiation* because there can be an instance of each resource and service available at each label. However, there is also a unification principle known as a single system image which is applied to the entire operating environment. All the zones are centrally administered from a special, protected global zone which manages the Trusted Computing Base (TCB) known as the Trusted Path. The zones share a single LDAP directory in which network-wide policy is defined, as well as a single name service cache daemon for synchronizing local databases. All labeling policy and account management is done from within the Trusted Path. MLS policy enforcement is automatic in labeled zones and applies to all their processes, even those running as root. Access to the Global Zone (and hence Trusted Path applications) is restricted to administrative roles.

8.3   TRUSTED EXTENSIONS MEDIATION

Trusted Extensions mediates access at the level of zones which results in a higher level of enforcement and fewer MLS access checks than a typical MLS operating system. For example, instead of maintaining labels on fine-grained objects like files and directories, Trusted Extensions associates labels with Solaris Containers (zones), and network endpoints. Each zone is assigned a unique sensitivity label and can be customized with its own set of file systems and network resources. Each mounted file system is automatically labeled by the kernel when it is mounted. The file system label is derived from the label of the zone or host which is sharing it. All files and directories within the mounted file system have the same label as their mount point. Because no explicit extensions to the file or file system structure are required, essentially any file system that works on Solaris 10 will work when Trusted Extensions label enforcement is enabled. This includes Sun issued file systems such as UFS, ZFS, SAM-FS and QFS as well as third-party file systems.

Processes are uniquely labeled according to the zone in which they are executing. All processes within a zone (and their descendants) must have the same label, and are completely isolated from processes in other zones. Unlike other virtualization technologies, there is no performance penalty for executing within a zoned environment as there is no emulation required for a Container. That is, system calls in all zones are handled by the Solaris kernel directly. Labeled zones can be instantiated quickly by cloning a copy of a default zone. Disk usage is minimized by sharing immutable instances of most system files and by utilizing copy-on-write technology for the rest.

A zone's local file systems are only writable at the zone's label, but can be shared with labeled zones via loopback or NFS mounts. Loopback mounts are used between zones running on the same host, and multilevel NFS is used for access between hosts. MLS protections are enforced on the mounts, and some integrity protections are also provided. For example, file systems that are shared by all zones on a system are always mounted read-only. Such file systems are assigned the lowest administrative label, admin_low. Similarly, file systems imported from lower-level zones are assigned the label of that lower-level zone with which they are shared.

File sharing between sets of Trusted Extensions systems using NFS can be symmetric. Corresponding zones on each system with matching labels can have read-write access to each other's shared file systems. Zones which dominate (i.e., have higher labels) than the owning zone can be granted read-only access depending on per-zone policy settings.

Writing up to higher-level regular files is not possible because such files are never visible within a labeled zone. However, writing up is possible using named pipes which are loopback mounted into higher-level zones. This unidirectional conduit is useful for implementing one-way guards and for tamper-proof logging.

Example 8.1. Figure 8.1 shows an example of the labels assigned to the mount points in a zone called needtoknow, whose label is CONFIDENTIAL : NEED TO KNOW. It dominates two user zones, internal and public. Zone needtoknow has read-write access to file systems mounted in its own zone, but read-only access to the dominated, user zones.

Mount Point	Access	Sensitivity Label
/	Read/Write	CONFIDENTIAL : NEED TO KNOW
/kernel	Read Only	ADMIN_LOW
/lib	Read Only	ADMIN_LOW
/opt	Read Only	ADMIN_LOW
/platform	Read Only	ADMIN_LOW
/sbin	Read Only	ADMIN_LOW
/usr	Read Only	ADMIN_LOW
/var/tsol/doors	Read Only	ADMIN_LOW
/tmp	Read/Write	CONFIDENTIAL : NEED TO KNOW
/var/run	Read/Write	CONFIDENTIAL : NEED TO KNOW
/home/jdoe	Read/Write	CONFIDENTIAL : NEED TO KNOW
/zone/public/export/home/jdoe	Read Only	PUBLIC
/zone/internal/export/home/jdoe	Read Only	CONFIDENTIAL : INTERNAL USE ONLY

Figure 8.1: Labeled mount attributes for Trusted Extensions file systems in Example 8.1.

To prevent configuration errors and to simplify system administration, there are no interfaces for specifying the labels of mount points. Instead, the kernel determines the labels of all mount points based on host and zone labels, and ensures that the MLS policy is correctly implemented.

By default, each labeled zone is completely isolated from all other labeled zones because their labels are required to be unique. No process in a zone can view or signal processes running in other zones. There are no privileges available for any process in a labeled zone to write to lower-level files. However, such policies as reading from files in lower-level zones, exporting directories to higher-level zones, and moving files into higher level zones can be enabled by specifying the privileges available to each zone when it is booted [1]. Privileges available to a zone can, in turn, be assigned to processes in the zone. However, a zone's privilege limit is an upper bound that applies to all processes (even root-owned) that are run in the zone. All policies that affect multiple zones, such as sharing of directories, are administered via the Trusted Path.

If multiple users are running processes at the same MLS label, these processes run in the same zone and are controlled using traditional DAC policies within the zone. Of course, the communications that these processes can make are controlled by the MLS policies on the zone as described above.

8.4   PROCESS RIGHTS MANAGEMENT (PRIVILEGES)

The Solaris operating system implements a set of privileges that provide fine-grained control over the actions of processes. Traditionally, Unix-based systems have relied on the concept of a specially-identified, super-user, called `root`. This concept of a Unix super-user has been replaced in a backward compatible manner with the ability to grant one or more specific privileges that enable processes to perform otherwise restricted operations.

The privilege-based security model is equally applicable to processes running under user id 0 (`root`) or under any other user id. For root-owned processes, the ability to access and modify critical system resources is restricted by removing privileges from these processes. For user-owned processes privileges are added to explicitly allow them to access such critical resources. The implications for such privilege-aware processes are both, that `root` processes can run more safely because their powers are limited, and that many processes that formerly required to be `root` processes can now be executed by regular users by simply giving them the additionally required privileges. Experience with modifying a large set of Solaris programs to be privilege-aware revealed an interesting fact. Most programs that formerly required to be executed as user `root` require only very few additional privileges and in many cases require them only once before they can be relinquished.

The change to a primarily privilege-based security model in the Solaris operating system gives developers an opportunity to restrict processes to those privileged operations actually needed instead of having to choose between all (super-user) or no privileges (non-zero UIDs). Additionally, a set of previously unrestricted operations now require a privilege; these privileges are dubbed the "basic" privileges. These are privileges that used to be always available to unprivileged processes. By default, processes still have the basic privileges.

---

[1] The policy for reading down is configurable because it is not always appropriate. For example, it could result in processes at a higher level executing lower-level applications which manipulate the higher-level data. One way to mitigate that risk is to configure the zone without the privilege to do read down mounts. An additional way is to specify the noexec mount option for lower-level mounts.

The single, all-powerful UID 0 assigned to a root process has been replaced with at least 68 discrete privileges that can be individually assigned to processes using the Service Management Facility (SMF), Role-based Access Control (RBAC), or a command-line program, such as ppriv(1).

Taken together, all defined privileges with the exception of the "basic" privileges compose the set of privileges that are traditionally associated with the root user. The "basic" privileges are "privileges" unprivileged processes were accustomed to having.

The privilege implementation in Solaris extends the process credential with four privilege sets:

- **I, the inheritable set**: The privileges inherited on exec.

- **P, the permitted set**: The maximum set of privileges for the process.

- **E, the effective set**: The privileges currently in effect.

- **L, the limit set**: The upper bound of the privileges a process and its offspring can obtain. Changes to L take effect on the next exec.

As shown in Figure 8.2, the sets I, P and E are typically identical to the basic set of privileges for unprivileged processes. The limit set is typically the full set of privileges.

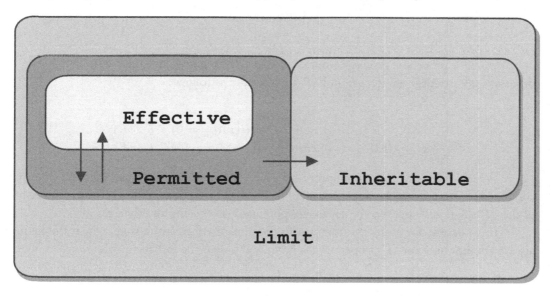

Figure 8.2: The relationship among Solaris privilege sets.

8.4.1 PRIVILEGE BRACKETING AND RELINQUISHING

The implementation of Solaris privileges empowers application developers to control how privileges are used within their programs. Using a technique called *privilege bracketing*, developers can write their programs such that they are only running with privileges when they are needed. Even more importantly, programs can not only enable or disable their privileges, but they can also drop any privileges granted to them (assuming they will not be needed) and even relinquish them (so they can no longer be used) when there is no longer a need for the privilege. Just as importantly, programs can also restrict which of their privileges can be passed along to their children (e.g., programs that they execute). In the Solaris operating environment many setuid programs (e.g., ping, traceroute, rmformat) and system services (e.g., nfsd, ftpd, mountd) use these techniques.

Each process has a Privilege Awareness State (PAS) that can take the value PA (privilege-aware) and NPA (not privilege-aware). PAS is a transitional mechanism that allows a choice between full compatibility with the old superuser model and completely ignoring the effective UID. To facilitate the discussion, we introduce the notion of "observed effective set" (oE) and "observed permitted set" (oP) and the implementation sets iE and iP.

A process becomes privilege-aware either by manipulating the effective, permitted, or limit privilege sets through the `setppriv` or `setpflags` system calls. In all cases, oE and oP are invariant in the process of becoming privilege-aware. In the process of becoming privilege-aware, the following assignments take place:

$$iE = oE \tag{8.1}$$
$$iP = oP \tag{8.2}$$

When a process is privilege-aware, oE and oP are invariant under UID changes. When a process is not privilege-aware, oE and oP are observed as follows:

$$oE = ((euid == 0) \ ? \ L : iE) \tag{8.3}$$
$$oP = ((euid == 0 || ruid == 0 || suid == 0)) \ ? \ L : iP) \tag{8.4}$$

When a non-privilege-aware process has an effective UID of 0, it can exercise the privileges contained in its limit set, the upper bound of its privileges. If a non-privilege-aware process has any of the UIDs 0, it will appear to be capable of potentially exercising all privileges in L.

It is possible for a process to return to the non-privilege aware state, which the kernel will always attempt on `exec`. This operation is permitted only if the following conditions are met:

- If any of the UIDs is equal to 0, P must be equal to L.

- If the effective UID is equal to 0, E must be equal to L.

When a process gives up privilege awareness, the following assignments take place:

$$if(euid == 0) \ iE = L\&I \tag{8.5}$$
$$if(anyuid == 0) \ iP = L\&I \tag{8.6}$$

The processes that do not have a UID of 0 will be assigned the inheritable set of privileges from its parent, as restricted by the limit set. When executing, the privileges in the process's (observed) effective privilege set permit the process to perform restricted operations. A process can use any of the privilege manipulation functions to add or remove privileges from the privilege sets. Privileges can be removed always. Only privileges found in the permitted set can be added to the effective and inheritable set. The limit set cannot grow. Note in Figure 8.2 that the inheritable set can be larger than the permitted set. Thus, it is possible that a process's children may have permissions additional to the process's permitted set.

When a process performs an exec, the kernel will first try to relinquish privilege awareness before making the following privilege set modifications:

$$E' = P' = I' = L\&I \tag{8.7}$$
$$L \ is \ unchanged \tag{8.8}$$

If a process has not manipulated its privileges, the privilege sets effectively remain the same, as E, P and I are already identical.

The limit set is enforced at exec time.

To run a non-privilege-aware application in a backward-compatible manner, a privilege-aware application should start the non-privilege-aware application with $I = basic$.

For most privileges, absence of the privilege simply results in a failure. In some instances, the absence of a privilege can cause system calls to behave differently. In other instances, the removal of a privilege can force a setuid application to seriously malfunction. Privileges of this type are considered "unsafe". When a process is lacking any of the unsafe privileges from its limit set, the system will not honor the setuid bit of setuid root applications.

8.4.2   CONTROLLING PRIVILEGE ESCALATION

In certain circumstances, a single privilege could lead to a process gaining one or more additional privileges that were not explicitly granted to that process. To prevent such an escalation of privileges, the security policy will require explicit permission for those additional privileges.

Common examples of escalation are those mechanisms that allow modification of system resources through "raw" interfaces; for example, changing kernel data structures through /dev/kmem or changing files through /dev/dsk/*. A special case of this is manipulating or creating objects owned by UID 0 or trying to obtain UID 0 using the setuid system call. The special treatment of UID 0 is needed because the UID 0 owns all system configuration files and ordinary file protection mechanisms allow only processes with UID 0 to modify the system configuration. While with appropriate file modifications, a given process running with an effective UID of 0 could gain all privileges, other protection mechanisms come into play.

When a process needs the permissions only available to root, it must be run with UID 0. Ultimately, we would like to eliminate this requirement for an all-powerful root principal, so that processes could be simply bestowed their appropriate privilege sets. However, better protection of the file resources above is necessary to prevent a malicious process from circumventing such limitations.

Of course, administrators should use as few UID 0 processes as possible. This reduces the size of the system's trusted computing base, thus limiting the number of processes that must be tamperproof. Where possible, a `root` process should be replaced with programs running under a different UID but with exactly the privileges they need. For example, daemons that never need to `exec` subprocesses should remove the privilege to execute processes from their permitted and limit sets.

8.4.3   ASSIGNED PRIVILEGES AND SAFEGUARDS

While it is possible for privileges to be assigned to a user, they should really be assigned to programs. An system administrator could give a user more powers than intended. The administrator should consider whether additional safeguards are needed. For example, if the privilege to lock process memory is given to a user, the administrator should consider setting the `project.max-locked-memory` resource control as well, to prevent that user from locking all memory.

8.5   ROLE-BASED ACCESS CONTROL (RBAC)

Role-based Access Control (RBAC) in Solaris is an alternative to the all-or-nothing security model of traditional superuser-based systems. With RBAC [94], an administrator can assign privileged functions to specific user accounts (or special accounts, called *roles*). RBAC is in keeping with the security principle of least privilege by allowing organizations to selectively grant privileges to users or roles based upon their unique needs and requirements. In general, organizations are strongly encouraged to use Solaris RBAC to restrict access to privileged operations rather than granting users complete access to the backwardly compatible root account.

Solaris RBAC was introduced in the Solaris 8 operating system, having come from Trusted Solaris, and has been enhanced and expanded with each new release of Solaris. Solaris RBAC functionality contains several discrete elements that can be used individually or together including authorizations, privileges, rights profiles and role designations. Figure 8.3 illustrates the relationship between these elements that are described in turn below.

8.5.1   RBAC AUTHORIZATIONS

An authorization is a unique string that represents a user's right to perform some operation or class of operations. Authorization definitions are stored in a database called `auth_attr`. For programming authorization checks, only the authorization name is significant.

Some typical values in an `auth_attr` database are shown below.

```
solaris.jobs.::::Cron and At Jobs::help=JobHeader.html
solaris.jobs.grant:::Delegate Cron & At \
   Administration::help=JobsGrant.html
```

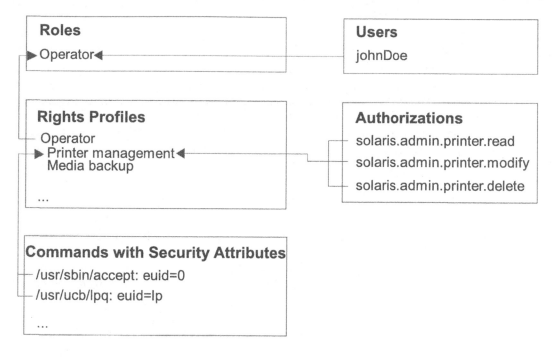

Figure 8.3: Relationship between RBAC elements.

```
solaris.jobs.admin:::Manage All Jobs::help=AuthJobsAdmin.html
solaris.jobs.user:::Cron & At User::help=JobsUser.html
```

Authorization name strings ending with the grant suffix are special authorizations that give a user the ability to delegate authorizations with the same prefix and functional area to other users.

An authorization is a permission that can be assigned to a role, be assigned to a user, or be embedded in a rights profile for performing a class of actions that are otherwise prohibited by security policy. Very often, authorizations are used in concert with privileged programs or services for the purpose of access control. For example, access to crontab is denied:

- If /etc/cron.d/cron.allow exists and the user's name is not in it.

- If /etc/cron.d/cron.allow does not exist and user's name is in /etc/cron.d/cron.deny.

- If neither file exists, only a user with the solaris.jobs.user authorization is allowed to submit a job.

In this case, the solaris.jobs.user authorization can be used to grant access to the cron facility (when other access control mechanisms are not present).

8.5.2   RIGHTS PROFILES

A *rights profile* is a collection of overrides that can be assigned to a role or user. A rights profile can consist of authorizations, individual commands, and other rights profiles. Each of the commands stored in a rights profile can define security attributes that determine how the program will be run. The following is the list of security attributes that can be assigned to commands in a rights profile:

- **uid (euid)**: The euid and uid attributes contain a single user name or a numeric user ID. Commands designated with euid run with the effective UID indicated, which is similar to setting the setuid bit on an executable file. Commands designated with uid run with both the real and effective UIDs.

- **gid (egid)**: The egid and gid attributes contain a single group name or a numeric group ID. Commands designated with egid run with the effective GID indicated, which is similar to setting the setgid bit on a file. Commands designated with gid run with both the real and effective GIDs.

- **privs**. The privs attribute contains a privilege set which will be added to the inheritable set prior to running the command.

- **limitprivs**. The limitprivs attribute contains a privilege set which will be assigned to the limit set prior to running the command.

8.5.3   USERS AND ROLES

A Solaris role is a special identity for running privileged applications that can be assumed by assigned users only. A role is similar to a normal user in that it has its own UID, GID, home directory, shell and password. A role differs from a normal user in two ways:

- A role cannot be used to (initially) log directly into a system either at the console or by any remote access service. Users must first log into the system before assuming a role.

- A role can only be accessed by a user who has previously been authorized to assume that role.

   Most often, roles are used for administrative accounts to restrict access to sensitive operations as well as for service accounts (e.g., web server or application server UID). It is important to ensure that actions taken by such accounts be attributable back to a specific user (who accessed the role). It should also be noted that delayed jobs (e.g., cron or batch) are independent of role assumption.

8.5.4   CONVERTING THE SUPERUSER TO A ROLE

Taken together, authorizations, rights profiles and roles offer the ability to delegate access to administrative functions with a level of detail that can be customized based upon an organization's policies and requirements. One of the most commonly cited examples of RBAC is the conversion of the root account to a role.

By implementing this change, `root` no longer will be able to directly log into the system, and `root` will only be able to be accessed by those possessing the correct credentials and explicit approval to assume that role. It is critical therefore that at least one user account be assigned to the `root` role, otherwise the role itself would no longer be able to be accessed. Note that the risk of administrators being unable to log in and assume the `root` role to perform privileged operations can be reduced by ensuring that their accounts have account lockout disabled, are stored in the local ("files") password tables, and have home directories that are mounted locally rather than over NFS. Solaris can still be configured such that booting the Solaris system into single user mode will enable administrators to log into the system directly as `root`, thereby providing a worst-case mechanism to access a privileged shell.

In addition, there are a number of other rights profiles provided in the Solaris OS by default including:

- **Primary Administrator**. Provides all of the capabilities of "superuser" in one profile. This profile grants rights that are equivalent to root.

- **System Administrator**. Provides a profile that can do most of the "superuser" tasks but fewer connected with security administration. For example, this role can create accounts but it cannot set or reset user passwords.

- **Operator**. Provides limited capabilities to manage files and offline media.

Such profiles define sets of rights associated with a particular job, as is a common use of role-based access control.

8.6   TRUSTED EXTENSIONS NETWORKING

A key feature in Trusted Extensions is its labeled networking that enables distributed computation to be controlled relative to the MLS policy. As in previous versions of Trusted Extensions software, remote hosts can be single-level or multilevel. Single level hosts have an implicit label assigned to them based on their network or IP address. Nonlabel aware systems, such as workstations running Microsoft Windows (TM), are assigned a specific label for communications purposes. Multilevel hosts are trusted to operate at a range of labels, and explicitly specify the label of every network packet when communicating with other trusted systems. Packet labels are specified using the Commercial IP Security Option (CIPSO) which encapsulates a sensitivity label as an IP option [53]. CIPSO is specified in the FIPS 188 Standard and is supported by Trusted Solaris 8 and other labeled systems.

When specifying the labeling policy for network attributes, both label ranges and sets of disjoint labels can be enumerated. This ability to precisely define the labeling policy is required to support various multilevel configurations including guards, NFS servers, Sun Ray servers, name servers, print servers, workstations, and high-assurance grid computing. An administrator can also assign a label range to a router even if the router does not interpret labels. Although zones have unique labels, specific multilevel services can be configured for each zone.

The network attributes database is maintained in an LDAP directory and shared by all trusted systems comprising a network of multilevel systems. IPsec can be used to authenticate the source IP addresses associated with incoming network packets. IPsec enforces integrity protection, and is used to encrypt data on multilevel networks.

Zones can be configured to share a single IP address, or they can be assigned unique IP addresses. Similarly, they can share the same physical network interface, or can be configured to use separate network interfaces. Both shared and per-zone IP addresses can be used concurrently, with different labeling policies for each IP address. Solaris Zones technology allows multiple zones to share a single network interface through the use of virtual interfaces.

Sharing of IP addresses is possible in Trusted Extensions because each packet is labeled. When a packet is received, the kernel uses the label of the packet to determine the appropriate zone to which it is authorized to be delivered. Sharing a single IP address for all zones is convenient for workstations and laptops, especially when DHCP is used. This simplifies deployment into infrastructures with limited IP addresses.

8.7   TRUSTED EXTENSIONS MULTILEVEL SERVICES

By default Solaris 10 with Trusted Extensions enables the following multilevel services:

- X11 Window System with the Common Desktop Environment (CDE) or the Gnome-Desktop.

- Printing using the Internet Protocol Printing or BSD Protocol Printing

- Network File System

- Sun Directory Server (LDAP server)

- Label Translation Service

- Name Service Cache Daemon

All other services are polyinstantiated in each zone. However, additional multilevel services such as Web Servers and Secure Shell can be enabled administratively via the Trusted Path. We discuss the multilevel window system and printing in detail below. We also discuss the use of multilevel services across the network, using the labeled networking described in the previous section.

Users can log in via the Trusted Path and can be authorized to select their multilevel desktop preference (CDE or Gnome-based). Once authenticated they are presented with an option to select an explicit label or a range of labels within their clearance and the label range of their workstation or Sun Ray desktop unit. The window system initiates a user session in the zone whose label corresponds to the user's default or minimum label.

The window system provides menus for interacting with the Trusted Path to change the label of the current workspace or to create additional labeled workspaces. For each selected label, the window system starts another user session in the corresponding zone. All of these user sessions run concurrently and are subjects of the user's identity that was established during the initial authentication. Each window is visibly labeled according to the zone or host with which it is associated. Although users can simultaneously interact with windows running in multiple zones, the applications themselves remain isolated.

Attempts to cut and paste data, or drag and drop files between clients running in different zone are mediated by the Trusted Path. Specific authorizations are required for upgrading or downgrading selections and files, and are prohibited by default. Figure 8.4 shows a screen shot of an authorized user interacting with the Trusted Path to upgrade a selection.

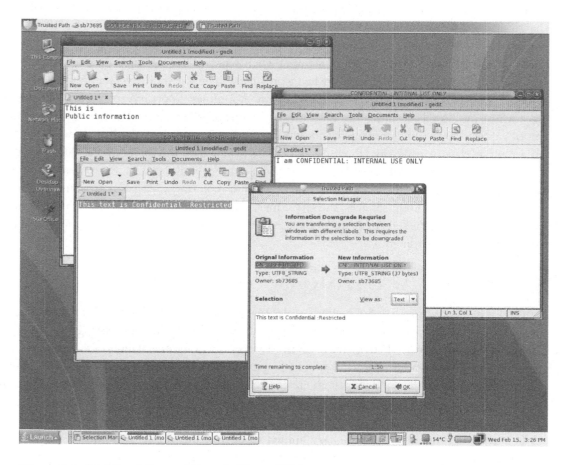

Figure 8.4: Multilevel Cut and Paste in Trusted JDS

Devices represent a security threat because they can be used to import and export data from the system. In Trusted Extensions, removable media devices are administered through the Trusted Path menu. The window system provides a Trusted Path interface for device allocation which provides fine-grained access to specific devices based on user authorizations and label ranges. For example, a user can be authorized to allocate the audio system (speaker and microphone) at a single level. Hot pluggable devices such as USB flash memory drives are also managed by the Trusted Path user interface. An authorized user can request to have such devices mounted into a zone whose label is within the user's label range and the device's label range. As an extra security measure, the raw device is not available within the labeled zone. This capability protects the integrity of the mounted file system and prevents unauthorized access.

Each printer is assigned a label range from which it will accept requests. Multilevel printers can accept jobs from labeled zones or remote hosts whose labels fall within their range. Each job can be encapsulated between reliably matching banner and trailer pages which indicate the label and handling caveats for the output. Each page can be automatically labeled with headers and footers corresponding to the sensitivity of the data.

Using per-zone IP addresses is required when separate networks are in use, and might be appropriate when multilevel services are being provided. To enable multilevel services, a database of multilevel ports is maintained via the Trusted Path. A multilevel port is a special kind of reserved port whose multilevel semantics are administratively controlled. For each IP address, a range and/or set of explicit labels can be configured for use by multilevel services. A privileged server can bind to a multilevel port using any IP addresses that are assigned to the server's zone. The server can receive requests at these labels and reply to any request. For multilevel TCP services, the reply is automatically sent using the label of the request without requiring any special programming in the server. For multilevel UDP services, the server must set a socket option to indicate the label of the reply. In either case, the server can query the kernel to determine the label of each request and then restrict the reply accordingly.

8.8   TRUSTED EXTENSIONS ADMINISTRATION

Trusted Extensions provides administrative roles that permit authorized users with the permissions to configure the system's security and an auditing subsystem for tracking processing.

By using the Trusted Path menu, authorized users can assume one or more of the administrative roles which they have been assigned. For obtaining each role, a secondary authentication is required. Once authenticated, the window system creates a new administrative workspace for the role and starts another session. These administrative workspaces are protected from interference by untrusted X11 clients and nonrole user logins. For roles who are cleared for all labels, their sessions are initiated as Trusted Path processes. Each role has a limited set of Role Based Access Control (RBAC) rights which restrict its access. Typically, two or more cooperating roles can be used to configure the system. For example, a system administrator role creates accounts and zones, while a security administrator assigns labels to them. Roles with sufficient rights can configure aspects of the MLS and DAC policies that apply to one or more zones.

Trusted Extensions audit records are compatible with standard Solaris OS. They include the labels of subjects and objects, and additional label-related events. The auditing system is configured via the Trusted Path and is transparent to users and roles running in labeled zones. The auditing system is robust and cannot be tampered with by processes running in labeled zones. Even processes with all privileges in a labeled zone cannot observe the audit trail nor tamper with any records.

8.9   SUMMARY

Solaris Trusted Extensions builds on the security features in Solaris 10 in an upward compatible fashion. Administrators of Solaris 10 can enable Trusted Extensions by turning the labeling service on with a single SMF command. Because it is now integrated with the Solaris OS, all of the latest Solaris functionality is supported by Trusted Extensions, and all hardware platforms are supported. Conversely, privileges and RBAC are fully supported in regular Solaris without the need to make use of the MLS features of Trusted Extensions. The MLS policy is enforced at the zone boundaries, rather than on individual processes or files, so access control is coarser-grained than traditional MLS operating systems. Note that this approach is similar to that envisioned for virtual machine systems, see Chapter 11. User applications running within zones require no customization for the Trusted Extensions platform. Trusted Extensions provides multilevel services that provide general services for applications at all levels without requiring application modifications.

CHAPTER 9

Case Study: Building a Secure Operating System for Linux

The Linux operating system is a complete reimplementation of the POSIX interface initiated by Linus Torvalds [187]. Linux gained popularity throughout the 1990s, resulting in the promotion of Linux as a viable alternative to Windows, particularly for server systems (e.g., web servers). As Linux achieved acceptance, variety of efforts began to address the security problems of traditional UNIX systems (see Chapter 4). In this chapter, we describe the resulting approach for enforcing mandatory access control, the Linux Security Modules (LSM) framework. The LSM framework defines a reference monitor interface for Linux behind which a variety of reference monitor implementations are possible. We also examine one of the LSM reference monitors, Security-enhanced Linux (SELinux), and evaluate how it uses the LSM framework to implement the reference monitor guarantees of Chapter 2.

9.1 LINUX SECURITY MODULES

The Linux Security Modules (LSM) framework is a reference monitor system for the Linux kernel [342]. The LSM framework consists of two parts, a reference monitor interface integrated into the mainline Linux kernel (since version 2.6) and a reference monitor module, called an LSM, that implements reference monitor function (i.e., authorization module and policy store, see Chapter 2 for the reference monitor concept definition) behind that interface. Several independent modules have been developed for the LSM framework [228, 183, 127, 229] to implement different forms of reference monitor function (e.g., different policy modules). We examine the LSM reference monitor interface in this section, and one of its LSMs, Security-Enhanced Linux (SELinux) [229] in the subsequent section. There are other LSMs and a debate remains over which LSM approach is most effective, but SELinux is certainly the most comprehensive of the LSMs.

9.1.1 LSM HISTORY

In the late 1990s, the Linux operating system gained the necessary support to make it a viable alternative in the UNIX system market. Although there were a variety of UNIX variants, such as AIX and HP/UNIX, and even other open source systems, such as the BSD variants, Linux became the mindshare leader among UNIX systems. Large server vendors, such as IBM and HP, threw their support behind Linux, and it soon became the main competitor to Microsoft Windows.

Also, in the late 1990s, a number of projects emerged that retrofit various security features into the Linux kernel. Since Linux was open source, anyone could modify it to meet their requirements (as long as they released their code back to the community, per the GNU Public License requirements [112]). A variety of prototype systems emerged, including Argus PitBull [13], LIDS [183], AppArmor (originally called Subdomain) [228], RSBAC [240], GRSecurity [296], DTE (see Chapter 7) for Linux [127], Medusa DS9 [204], OpenWall [236], HP's Secure OS Software for Linux [80], and a retrofit of the former Flask/DTOS/DTMach system (see Chapter 7), now called SELinux. All these modified Linux systems varied in fundamental ways, but all aimed to provide a valuable security function. AppArmor and PitBull were both sold as commercial products.

In 2001, momentum was growing for inclusion of a reference monitor in the Linux kernel. Problems with worms, viruses, and denial-of-service attacks were reaching a significant level, although mostly on the Windows platform. At the Linux kernel summit that year, the SELinux prototype was presented, and the Linux community, including Linus Torvalds in particular, seemed to accept the idea that a reference monitor was necessary. However, Linus faced two challenges. First, he was not a security expert, so he could not easily decide among the approaches and felt it was not appropriate for him to make such a decision. Second, the security community itself could not agree on a single, "best" approach, so Linus could not depend on the security community to guide him to a single approach. As a result, Linus argued for a design based on kernel modules where a single interface could support all the necessary modules. This approach became the LSM framework.

A community formed around the idea of building a single reference monitor interface for Linux (although not all the Linux security prototype researchers agreed [239, 297] [1]), and this community designed and implemented the LSM framework. The main task was to implement the LSM reference monitor interface. The design of the LSM framework's reference monitor interface had the following goals [342]:

- The reference monitor interface must be truly generic, such that "using a different security model is merely matter of loading a different kernel module"

- The reference monitor interfaces must be "conceptually simple, minimally invasive, and efficient"

- Must support the POSIX.1e capabilities mechanism as an "optional security module"

The first two requirements motivated collecting the union of the authorization queries from all previous Linux security, such that all modules could be supported, but restricting the number of authorization queries as much as possible to prevent redundant authorizations that made add complexity and impact performance. While the LSM interface was designed manually [342], source code analysis tools were built to verify the completeness [351] and consistency [149] of the LSM interface, finding six interface bugs that were resolved.

Performance analysis showed that the most of the LSM interface had no tangible performance impact [342], but the CIPSO implementation (i.e., labeled networking, see Section 7.5.2) provided

---

[1]The RSBAC comment dated April 2006 that LSM would be removed from the official kernel is no longer current. Linus Torvalds reaffirmed his support for LSM in the 2006 Linux Kernel Summit, and LSM will remain part of the mainline Linux kernel for the foreseeable future.

with the initial LSM interface was rejected. The performance overhead of keeping labels consistent under packet fragmentation and defragmentation, even if no security policy was being enforced, was too costly. Two other alternatives for labeled networking are now supported by the Linux kernel. First, Labeled IPsec [148], based on the Flask labeled networking [50], negotiates labels for IPsec network communications in addition to cryptographic parameters. An LSM controls network communication by authorizing whether a process can use a particular IPsec communication channel. Since the label of the IPsec channel is established at negotiation time, there is no need to include the label in the packet. Second, Paul Moore built a new implementation of CIPSO for Linux, called Netlabel [214]. Netlabel provides a less intrusive version of CIPSO which was accepted by the Linux community.

The Linux Security Modules framework was officially added to the Linux kernel with the release of version 2.6. The SELinux module and a module for implementing POSIX capabilities [307] were included with the release of LSM. Novell, the distributor of SuSE Linux, purchased the company that supported AppArmor, so SuSE and other Linux distributions support the AppArmor LSM as well.

SELinux and AppArmor have become the major LSMs. While both provide tangible Linux security improvements, converting Linux (or any UNIX system) to a system that can satisfy reference monitor guarantees is a difficult task. However, with Linus reaffirming his support for the LSM framework [188] in 2006 and a variety of Linux vendors support security behind LSMs, the LSM framework can be considered a modest success.

9.1.2   LSM IMPLEMENTATION

The LSM framework implementation consists of three parts: (1) the reference monitor interface definition; (2) the reference monitor interface placement; and (3) the reference monitor implementations themselves.

LSM Reference Monitor Interface Definition   The LSM interface definition specifies the ways that the Linux kernel can invoke the LSM reference monitor. Linux header file include/linux/security.h lists a set of function pointers that invoke functions in the loaded LSM. A single structure, called security_operations, contains all these LSM function pointers. We refer to these function pointers collectively as the *LSM hooks*. Fundamentally, the LSM hooks correspond to LSM authorization queries, but the LSM interface must also include LSM hooks for other LSM tasks, such as labeling, transition, and maintenance of labels.

Two examples of LSM hooks are shown below.

```
static inline int security_inode_create (struct inode *dir,
                                          struct dentry *dentry,
                                          int mode)
{
        if (unlikely (IS_PRIVATE (dir)))
                return 0;
```

```
        return security_ops->inode_create (dir, dentry, mode);
}

static inline int security_file_fcntl (struct file *file, unsigned int cmd,
                                        unsigned long arg)
{
        return security_ops->file_fcntl (file, cmd, arg);
}
```

First, `security_inode_create` authorizes whether a process is permitted by the LSM to create a new file, indicated by `dentry`, in a particular directory `dir`. The LSM hook is invoked through the call to the function pointer defined by `security_ops->inode_create`. The LSM loaded defines how the authorization is performed. Second, `security_file_fcntl` authorizes a specified process's ability to invoke `fcntl` on a specific file. Subsequent LSM hooks in the function `do_fcntl` enable an LSM to limit certain uses of `fcntl` independently (e.g., setting the `fowner` field that signals the associated process on file operations).

In all there are over 150 LSM hooks that enable authorizations (as above), and the other LSM operations of labeling, transitioning labels, and maintenance of labels. While different LSM hooks are intended to serve different purposes, they all have a similar format to the two listed above.

LSM Reference Monitor Interface Placement   The main challenge in the design of the LSM framework is the placement of the LSM hook. Most of the LSM hooks are associated with a specific system call, so for these the LSM hook is placed at the entry to the system call. However, several of the LSM hooks cannot be placed at the system call entry point (e.g., to prevent TOCTTOU attacks, see Chapter 2). For example, as shown in Figure 9.1, the open system call converts a file path to a file descriptor that enables access (i.e., read and/or write) to the associated file. Locating the specific file described by the file path requires authorizing access to the directories that accessed along the path, any file links in the path, and finally authorizing to the target file for the specific operations requested. Since these components are extracted from the file path at various points in the open processing, the LSM hook placement is nontrivial.

While there are some discretionary checks in place that guided the placement of the LSM hooks for open, the process by which the LSM hooks were placed was largely ad hoc. For ones where no previous discretionary authorization was performed, the implementors made a manual placement. Some placements were found to be wrong, and some security-sensitive operations were found to be lacking mediation, but these issues were resolved through source code analysis [351, 149].

An LSM hook is placed in the code using the inline function declarations (e.g., `security_inode_create`) which is expanded at compile-time to the LSM hooks as shown by the code above. Inline functions for each LSM hook are used to improve the readability of the code.

LSM Reference Monitor Implementations   Finally, LSMs must be built to perform the actual authorizations. Actual LSMs include AppArmor [228], the Linux Intrusion Detection System

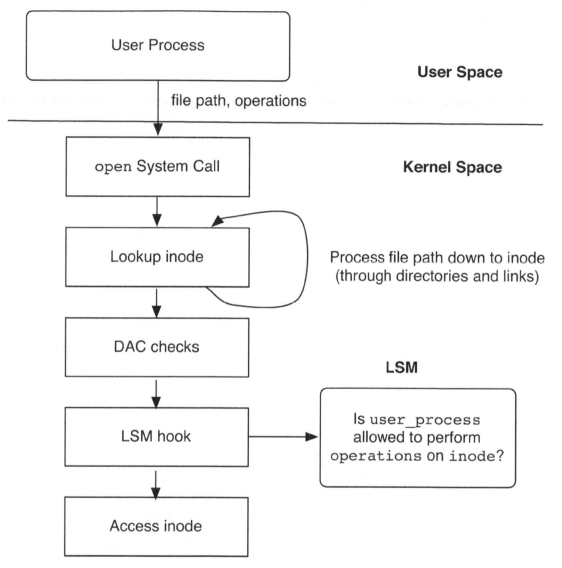

Figure 9.1: LSM Hook Architecture

(LIDS) [183], SELinux [229], and POSIX capabilities [307]. Each of LSMs provide a different approach to mandatory access control, excepting POSIX capabilities which was an existing discretionary mechanism in Linux. POSIX capabilities were converted to a module to enable independent development from the mainline kernel and because some LSMs aimed to implement the capability controls in a different manner [342].

Although Gasser and Schell identified that different security kernel policies require different reference monitor interfaces (i.e., different LSM hook placements) [10] (in the context of security kernels, see Chapter 6), the LSM uses the same placements for all LSMs. In practice, the choice of LSM hook placements was a union of the reference monitors being ported to the framework. The LSM implementation does not require that each LSM provide implementations for every hook, so the union approach does not demand extra work for LSM developers. Nonetheless, the set of LSM hooks has largely stabilized.

An example of the type of policy implemented by an LSM is the confinement policy of AppArmor [228]. AppArmor is a mandatory access control (MAC) system where the threat model is focused on the Internet. If we assume that systems are configured correctly, then the Internet is the only way that malicious input can reach the system. One threat is that network-facing daemons (e.g., `inetd`) are susceptible to malicious inputs (e.g., buffer overflows, format string vulnerabilities, etc.). AppArmor uses confinement policies for such network-facing daemons, that have traditionally been run with full privilege (e.g., `root`), to prevent compromised daemons from compromising the entire system.

9.2   SECURITY-ENHANCED LINUX

Security-Enhanced Linux (SELinux) is a system for enforcing mandatory access control that is based on the LSM framework [195, 229]. As shown in Figure 9.2, SELinux consists of a Linux Security Module and a set of trusted services for administration and secure system execution. In this section, we detail the SELinux reference monitor (Sections 9.2.1–9.2.4), then we discuss trusted services for administration (Section 9.2.5) and general trusted services (Section 9.2.6). We conclude this section with an evaluation of the SELinux system against a secure operating system specification in Definition 2.5.

The SELinux reference monitor consists of an authorization module and policy store. The SELinux authorization module builds authorization queries for a mandatory protection system (see Definition 2.4) in the SELinux policy store. SELinux uses fine-grained and flexible models for its protection state, labeling state, and transition state that cover all Linux system resources that are considered security-sensitive. Thus, the SELinux mandatory protection system enables comprehensive control of all processes, so policy writers can exactly define the required accesses. However, the low-level nature of the policy models results in complex policies that are difficult to relate to secrecy and integrity goals (e.g., information flow goals of Chapter 5). Nonetheless, the SELinux approach accurately demonstrates the challenges we face in ensuring that a commercial system enforces intended security goals.

9.2.1   SELINUX REFERENCE MONITOR

The SELinux reference monitor consists of two distinct processing steps. First, the SELinux reference monitor converts the input values from the LSM hooks into one or more authorization queries. These LSM hooks include references to Linux objects (e.g., file and socket object references), and

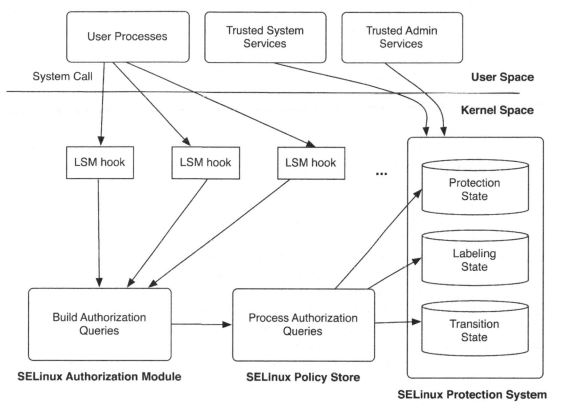

Figure 9.2: SELinux System Architecture

in some cases, argument flags, but the SELinux reference monitor must convert these into SELinux authorization queries (see below). Second, the core of the SELinux reference monitor processes these authorization queries against the SELinux protection system (i.e., the protection state, labeling state, and transition state). The SELinux protection system representation is highly optimized to support fine-grained authorization queries efficiently.

Consider the example below, the SELinux implementation behind the LSM hook that authorizes a file open system call.

```
static int selinux_inode_permission(struct inode *inode, int mask,
                                    struct nameidata *nd)
{
        if (!mask) {
                /* No permission to check. Existence test. */
                return 0;
        }
```

```
        return inode_has_perm(current, inode,
                              file_mask_to_av(inode->i_mode, mask), NULL);
}
```

Recall that when open system call is invoked, the target file is specified by a UNIX pathname and the requested operations are specified using a bit vector flags [2]. The kernel's implementation of open resolves the pathname down to the actual inode that refers to the target file, and then it invokes the LSM hook to authorize whether the requesting process can perform the requested operations on the resultant inode.

The function selinux_inode_permission above has three arguments, the inode for the file, the mask that indicates the file operations, and namedata related to the file path (not used in this authorization).

The SELinux implementation identifies the specific Linux objects corresponding to the subject, object, and operations for the authorization query. First, the subject of an open system call is the process that submitted the system call. In Linux, the process that invoked a system call is identified by the global variable current. As a result, this need not be an input from the LSM hook. Second, the object of an open call is the target inode. A reference to the inode is included in the LSM hook. Third, the operations requested by an open system call (e.g., read, write, and append) are determined from the flags input is sent to selinux_inode_permission function via the mask variable. The namedata is not used by the SELinux LSM, but may be used by other LSMs.

The subject (current), object (inode), and operations (results of the file_mask_to_av) are submitted to the function inode_has_perm, which generates the actual SELinux authorization query as shown below.

```
static int inode_has_perm(struct task_struct *tsk,
                          struct inode *inode,
                          u32 perms,
                          struct avc_audit_data *adp)
{
        struct task_security_struct *tsec;
        struct inode_security_struct *isec;
        struct avc_audit_data ad;

        tsec = tsk->security;
        isec = inode->i_security;

        if (!adp) {
                adp = &ad;
                AVC_AUDIT_DATA_INIT(&ad, FS);
                ad.u.fs.inode = inode;
        }
```

---

[2]The open system call has a third argument mode, but that is not pertinent to this example. Its implications are authorized elsewhere.

```
        return avc_has_perm(tsec->sid, isec->sid, isec->sclass, perms, adp);
}
```

Rather than submitting the objects directly in an authorization query, SELinux assigns labels to subjects and objects, called *contexts* in SELinux. As we describe in the next section, *subject contexts* define a set of permissions (objects and operations) available to processes running with that context. An *object context* groups a set of objects that have the same security requirements. As required for a mandatory protection system, the set of contexts must be fixed, so the protection state is immutable. Likewise, the labeling and transition states in an SELinux system are immutable as well.

In SELinux, the kernel stores a context with each process and system resource that may appear in an LSM hook. For processes, its data type `task_struct` includes the field `security` in which the subject context of the process is stored. For the `inode` data type, a field `i_security` stores its object context. The function `inode_has_perm` extracts the subject and object contexts for these input arguments (i.e., `tsk` and `inode` in `inode_has_perm`) and generates the SELinux authorization query, defined by the function `avc_has_perm`. This function takes four arguments: (1) the subject context; (2) the object context; (3) the SELinux classification for the object's data type; and (4) the operations requested in the query. The SELinux policy store executes this query, determining whether the subject context can perform the requested operations on an object with the specified object context and SELinux classification for its data type. Such classifications correspond to system data types, such as `file` and `socket`, as well as more specific subtypes of these.

SELinux defines authorization queries for nearly all of the LSM hooks. For most of these LSM hooks, a single authorization query is generated, but in some cases, multiple authorization queries are generated and evaluated. For example, in order to send a packet, the process must have access to send using the specified port, network interface, and IP address (see `selinux_sock_rcv_skb_compat` for an example). Each of these authorization queries must be authorized for the operation to be permitted.

Authorization queries on protection state retrieve the SELinux policy entry that corresponds to the subject context, object context and object data type. The policy entry contains the authorized operations for this combination, and `avc_has_perm` determines whether all the requested operations are permitted by the entry. If so, the operations are authorized and the SELinux implementation returns 0. The Linux kernel is then allowed to execute the remainder of the system call (or at least until the next LSM hook).

9.2.2   SELINUX PROTECTION STATE
The SELinux reference monitor enforces the SELinux protection state, labeling state, and transition state. First, we discuss the SELinux protection state. SELinux contexts described above represent the SELinux protection state. They are a rich representation of access control policy enabling the definition of fine-grained policies. In this section, we define the concepts in an SELinux context and describe how they express security requirements over Linux processes and system resources.

SELinux Contexts   Figure 9.3 shows the concepts that define an SELinux context and their relationships. A *user* is the SELinux concept that comes closest to a UNIX user identity (UID). The

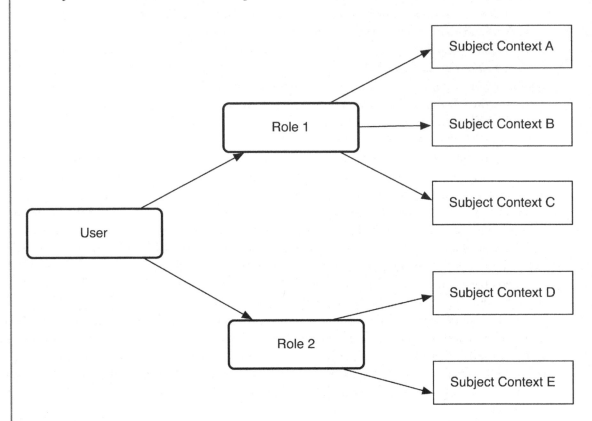

Figure 9.3: SELinux Contexts: *user* limits the set of *roles* that can be assumed to *Role 1* or *Role 2* (only one role). *Roles* limit the set of *subject types* and the *MLS range* that a process can assume (and hence, the permission available to the user). A context may have only one subject type, but a process can transition among all the subject types associated with a role (if authorized).

user is the authenticated identity of a user in the system. This typically corresponds to UNIX user identity in name, but does not convey any access rights to the user. In SELinux, user identity solely defines a set of roles to which the user is permitted by the SELinux policy. When a user authenticates (e.g., logs in), the user can choose one role from the set authorized by the SELinux policy.

An SELinux *role* is similar to the concept of a role in a role-based access control (RBAC) model [94, 268, 272] in that it limits the set of permissions that the user may access. However, unlike an RBAC role, the role is not assigned permissions directly. Rather, an SELinux role is associated with a set of *type* labels, as in a Type Enforcement (TE) model [33], and these type labels are assigned permissions. The role also optionally determines the *MLS range* that processes in that role may assume. When a user is authenticated, this determines the user's role and all the processes that the user runs must have a type label and MLS range that is authorized for the user's role.

For example, a user authenticates to the identity `alice` under the user role `user_r`. This role is permitted to run typical user processes under the type label `user_t`, but is also allowed to run other processes with type labels that `user_r` is authorized for. The user role `user_r` permits the user `alice` to use the `passwd` program to change her password, but since this program has different permissions than a typical user program (e.g., it can modify /etc/shadow, the file containing the hashed user passwords) it runs under a different type label, `passwd_t`. As a result, the user role `user_r` is authorized to run processes under the `user_t` and `passwd_t` type labels.

Type labels are assigned to permissions using `allow` statements as shown below.

```
allow <subject_type>  <object_type>:<object_class>  <operation_set>
allow user_t          passwd_exec_t:file            execute
allow passwd_t        shadow_t:file                 {read write}
```

An `allow` statement associates a subject type (e.g., `user_t`) with permissions described in terms of an object type (e.g., `passwd_exec_t`, the label of the `passwd` executable file), the data type of the object (e.g., the `passwd` executable is a file), and the set of operations on the object type authorized by the `allow` statement (e.g., `execute`). Thus, the first `allow` statement permits any process with the type label `user_t` to execute any file labeled with the `passwd_exec_t` type label. The second `allow` statement permits any process running with the `passwd_t` type label to read or write any file with the `shadow_t` type label. This limits access to /etc/shadow to only those processes that run under the `passwd_t` type label, typically only processes running `passwd`.

The SELinux MLS labels represent a traditional lattice policy consisting of *sensitivity levels* and *category sets*. The MLS labels are interpreted according to the Bell-LaPadula model [23] (see Chapter 5) for read operations, but a more conservative interpretation is used to authorize write operations. For a read operation, the subject's sensitivity level must dominate (i.e., be greater than) or be equal to the object's sensitivity level, and the subject's category sets must include all (i.e., be a superset of) the object's category sets (i.e., the *simple-security property*). For a write operation, the subject's sensitivity level must be equal to the object's sensitivity level, and the subjects category sets must equal those of the object as well. This is more restrictive than the usual MLS write rule based on the ⋆-security property. The ⋆-security property permits write-up, the ability for subjects in lower sensitivity levels to write to objects in higher sensitivity levels. The ⋆-property presumes that the lower secrecy processes do not know anything about the high secrecy files (i.e., the subject's sensitivity level is dominated by the object's), such as whether a higher secrecy file of a particular name exists. However, Linux processes are not so unpredictable, so it may be possible for one Linux process to guess the name of a file used by a higher secrecy process, thus impacting the integrity of the system. As a result, write-up is not permitted in SELinux.

The SELinux reference monitor authorizes whether the subject's type label and MLS label both permit the requested operations to be performed on the request object based on its type label and MLS label. The two labels are authorized independently as described above. For type labels, an `allow` rule must be defined that permits the subject type to perform the requested operations on the corresponding object type. In addition, the MLS labels of the subject and object must also permit the requested operation. Both authorization tests must pass before the operation is authorized.

SELinux Policies   The SELinux protection state enables comprehensive, fine-grained expression of a system's security requirements. First, each distinct object data type and operation in a Linux system is distinguished in the SELinux protection state. The SELinux protection state covers all security-sensitive system resource data types, including various types of files, various types of sockets, shared memory, interprocess communications, semaphores, etc. There are over 20 different object data types in all. In addition, SELinux provides a rich set of operations for each data type. In addition to the traditional `read`, `write`, and `execute` operations on a file, the standard file data type in SELinux has operations for `create`, `ioctl`, `fcntl`, extended attributes, etc. As a result, comprehensive and fine-grained control of system resources is possible.

Second, each process and object with different security requirements requires a distinct security context. If two processes cannot access exactly the same set of objects with exactly the same set of objects, then two distinct subject type labels are necessary, one for each process. Then, the appropriate `allow` statements for each can be defined. For example, distinct subject types for `user_t` and `passwd_t` had to be defined because `passwd` can access the `/etc/shadow` whereas a typical user process cannot. Further, if two objects cannot be accessed by exactly the same processes, then they also require two distinct object type labels. Again, the `shadow_t` and `passwd_exec_t` object type labels are necessary because these two files (`/etc/shadow` and the `passwd` executable) cannot be accessed by all processes using the same operations. As a result, over 1000 type labels are defined in the SELinux reference policy, the default SELinux protection state, and tens of thousands of `allow` statements are necessary to express all the different relationships between subjects and objects in a Linux system.

While the SELinux policy model results in complex protection state representations, the protection state complexity is a result of the complexity of Linux systems. Linux consists of many different programs, most with distinct access requirements and distinct security requirements, resulting in a large number of type labels. The large number of type labels then requires a large number of `allow` statements to express all the necessary access relationships. The SELinux reference policy demonstrates what we are up against in trying to build secure Linux systems.

9.2.3   SELINUX LABELING STATE

Since the SELinux protection state is defined in terms of labels, as is typical of a mandatory access policy, the protection state must be mapped to the actual system resources. We suspended some degree of disbelief in the last section because, while we mentioned that certain files, such as `/etc/shadow`, had certain labels, such as `shadow_t`, we did not specify how the files obtained these labels in the first place. Further, processes are also assigned labels, such as the `passwd` process having the `passwd_t` label, and the mapping of labels to processes must also be defined.

These specifications are provided in what we call the *labeling state* of the mandatory protection system in Definition 2.4. The labeling state is an immutable policy that defines how newly created processes and system resources are labeled. SELinux provides four ways in which an object's label can be defined.

First, an object may be labeled based on its location in the file system. Suppose the files /etc/passwd and /etc/shadow are provided in a Linux package for the passwd program. In this case, the file already exists in some form and needs to be labeled when it is installed. SELinux uses *file contexts* to label existing files or files provided in packages. A file context specification maps a file path expression to an object context. The file path expression is a regular expression that describes a set of files whose file path matches that expression. Below, we list two file contexts specifications.

```
<file path expr>   <context>
/etc/shadow.*      system_u:object_r:shadow_t:s0
/etc/*.*           system_u:object_r:etc_t:s0
```

For example, the second file context specification above defines the object context for files in the /etc directory. /etc/shadow gets a special context while other files in /etc (e.g., /etc/passwd) get the default context [3].

Second, for dynamically created objects, their labels are inherited from their parent object. For files, this is determined by the parent directory. For all files dynamically created in the /etc directory, they inherit the label defined for the directory, etc_t.

Third, type_transition rules can be specified in the SELinux policy that override the default object labeling. Below, we show a type_transition rule that relabels all files created by processes with the passwd_t type that would be assigned the etc_t label by default to the passwd_t label.

```
type_transition <creator_type> <default_type>:<class> <resultant_type>
type_transition passwd_t        etc_t:file                shadow_t
```

Note that the creating process context must be authorized to relabel these etc_t files to passwd_t files [4]. If we use the passwd process to create /etc/shadow, where /etc has the etc_t label, it would be assigned a shadow_t label instead based on this rule.

The SELinux labeling state enforces security goals through the administrator-specified file contexts, default labeling, and authorized type_transition rules. The labeling state enables precise control over labeling, but does not necessarily ensure a coherent security goal (i.e., information flow). An external analysis is necessary to determine whether labeling state achieves the desired security, as we discuss in the SELinux evaluation.

---

[3]Note that the user and role for all SELinux objects are system_u:object_r:.

[4]In order to relabel an object from type T1 to type T2, the subject must have allow rules that permit relabelfrom T1 and relabelto T2.

9.2.4   SELINUX TRANSITION STATE

By default, a process is labeled with the label of its parent, as described above, but the SELinux transition state enables process label transitions. If a user shell process runs with the `user_t` label, then all the processes that are created from this shell are also run under the `user_t` label. While this makes sense for many programs, such as editors, email client, and web browsers, some programs that the user runs need different permissions. For example, the `passwd` program needs access that must not be permitted for typical user programs, such as write access to `/etc/passwd` and read-write access to `/etc/shadow`.

SELinux `type_transition` rules are also used to express such process label transitions. As shown below the syntax is similar to the object labeling case, but the semantics are slightly different.

```
type_transition <current_type> <executable_file_type>:process <resultant_type>
type_transition user_t         passwd_exec_t:process              passwd_t
```

For process label transitions, a `type_transition` rule specifies that a process running in a specific label (i.e., the `current_type`) executes a file with a specific label (i.e., the `executable_file_type`), then the process is relabeled to the `resultant_type`.

As is the case for object labeling, process label transitions on execution must be authorized. This requires three SELinux permissions: (1) the process must have `execute` access to the executable file's type; (2) the process must be authorized to transition when executing that file; and (3) the process must be authorized to transition its label to the `resultant_type`. In the case above, the user shell forks itself and executes the `passwd` file. At execution time, the `type_transition` rule is invoked. The SELinux reference monitor retrieves such rules, and authorizes the conditions necessary to invoke the rule. If the transition is authorized, then the process is run using the `passwd_t` label, and it is able to access the `/etc/passwd` and `/etc/shadow` files as necessary.

Note that SELinux process label transitions are only permitted at process execution [5]. When a process is executed, the old process image is replaced with a new image defined by the file being executed, so the process context can be reassigned based on this image. Note that there may be some carryover from the old process, such as the set of file descriptors that are left open on execute and the processes environment variables, but the SELinux transition rules can limit the contexts under which transitions are allowed. For example, if a program depends on being run with a high integrity set of environment variables, then only transitions from high integrity contexts should be permitted. In the case of `passwd`, it is run from untrusted user processes, so the `passwd` executable must be trusted to protect itself from any low integrity inputs provided at execute time.

SELinux process transitions are more secure than traditional UNIX process transitions via `setuid` in several ways. First, a `setuid` transition almost always results in a process running with full system privileges (i.e., a `setuid root` process). In SELinux, the process transitions to a specific label with limited permissions defined for its purpose. Second, UNIX permits any process to execute a `setuid` program. As a result, all `setuid` programs are susceptible to malicious input from untrusted

---

[5] SELinux now has a command that permits process context transitions at any time, called `setcon`, but this command must be used carefully to prevent a process from obtaining unauthorized permissions. In general, use of this command is not recommended.

invocations. In SELinux, the contexts under which a process may be invoked can be limited. For example, SELinux rules can be written to ensure that only trusted contexts can execute a program and obtain all its rights.

SELinux transitions are comparable to Multics transitions, see Chapter 3. In Multics, ring brackets limit which processes may cause a process label transition by executing more trusted code. However, SELinux controls are finer-grained, as they can be defined at the level of an individual program, rather than a protection ring. However, Multics defines a formal concept for ensuring that a protection domain is protected from malicious inputs, the *gatekeepers*. SELinux has no such concept, but depends on the program developer to ensure protection.

Finally, SELinux now provides mechanisms that enable a process to relabel itself or system resources using the `setcon` and `chsid` commands, respectively. For system resources, the `passwd` process can explicitly invoke `chsid` to relabel `/etc/shadow` to the `shadow_t` label. Any process that is SELinux-aware can request a file relabeling, but the SELinux reference monitor authorizes all these transitions. That is, a `passwd_t` must also be authorized to relabel `passwd_t` files to `shadow_t`.

9.2.5   SELINUX ADMINISTRATION

As SELinux uses a mandatory access control (MAC) policy, only system administrators may modify its protection system's states. As a result, these states are generally static. SELinux provides two mechanisms for updating its protection system: (1) monolithic policy loading and (2) modular policy loading. In either case, configuring SELinux policies is a task for experts, so only a small number of policies have been developed.

Monolithic Policies   The traditional SELinux protection system states are defined as a single, comprehensive, binary representation generated from the policy statements (e.g., `allow`, `type_transition`, etc.) described above. The SELinux policy compiler `checkpolicy` builds such policy binaries. For a monolithic policy, the tens of thousands of SELinux policy statements are compiled into a binary that is over 3 MBs in size.

The trusted program `load_policy` enables an administrator to load a new protection state that entirely replaces the old protection state. `load_policy` uses the Linux Sysfs file system to load the binary into the Linux kernel where the SELinux reference monitor in the kernel can use it. All authorization queries are checked against the policy binary.

Modular Policies   As the SELinux policy is actually defined per Linux program and Linux programs themselves may be installed incrementally via packages, the SELinux policy administration mechanisms have also been extended to support incremental modification. SELinux policy modules define program-specific protection state contributions. A comprehensive SELinux policy binary is constructed from these individual modules. A policy module consists of four parts: (1) its own type labels and `allow` rules for these types; (2) its file context specification defining how its files are labeled; (3) its interfaces that enable other modules to access its type labels; and (4) this module's use of other module's interfaces.

The definition of type labels, `allow` rules, and file contexts are no different than for the monolithic policy, described in examples above, but the policy modules add the concept of module interfaces [324]. Module interfaces, like public method interfaces in object-oriented programs, provide entry points for other modules to access a module's type labels. For example, an interface definition specifies a set of `allow` rules that are permitted to the module invoking the interface. For example, the kernel policy module defines an interface `kernel_read_system_state(arg)` where the type label submitted as the argument `arg` is assigned to allow rules that permit read access to system state. A policy module specifies both its own interfaces and the set of interfaces that it uses. The function `semodule` is used to load new modules into the SELinux policy binary.

Policy Development   Originally, two types of SELinux policies were developed: (1) a strict policy and (2) a targeted policy. The strict policy aims to enforce least privilege over all Linux programs, thus maximizing the protection possible while permitting reasonable functionality. The strict policy presents two challenges to deployment. First, it may be more restrictive than the Linux programs expect, leading to the failure of some programs to run properly. Second, the strict policy does not enforce any formal secrecy or integrity goal, so the policy may still permit significant vulnerabilities.

The targeted policy principle was introduced by the AppArmor LSM [228], and it defines *least privilege* policies for network-facing daemons to protect the system from untrusted network input. Other programs run without restriction. This limits the task of configuring restrictive policies to just the network-facing daemons, which simplifies policy expression and debugging. However, the targeted policy does not protect the system from other low integrity inputs (e.g., malicious emails, downloaded code, malware that is installed under a different label). As a result, the targeted policy is more appropriate for server systems whose software is carefully managed, but which may be susceptible to malicious network requests. In practice, SELinux distributions (e.g., RedHat) are delivered with a *targeted* SELinux policy.

Recently, a third SELinux policy, the reference policy, has been defined [309]. The reference policy enables an administrator to build either the targeted or strict policy from a single set of policy files. A configuration file enables administrators to describe their specifications for building policies. The reference policy also includes MLS support by default.

9.2.6   SELINUX TRUSTED PROGRAMS

In addition to the administrator operations above to load policy (i.e., `load_policy` and `semodule`), there are many other user-level programs that are trusted to specify and/or enforce the SELinux security requirements for the SELinux system to be secure. These programs include authentication programs (e.g., `sshd`) necessary to establish an authenticated user's subject context, system services necessary to bootstrap the system (e.g., `init`), and server programs which are depended upon to enforce the SELinux policy on their operations.

Authentication programs have been modified to understand SELinux contexts. When a user authenticates, these programs inform the SELinux reference monitor, so it can assign the proper subject context for that user's processes.

System bootstrap services are mainly trusted because they have broad permissions that may enable them to compromise the integrity of the SELinux reference monitor and/or policy. These services run with near full privilege and are trusted not to modify or circumvent policy. For example, such processes use the traditional UNIX fork/exec when they start system services (e.g., `vsftpd`), so that these obtain the proper set of access rights through process labeling (i.e., via `type_transition` rules), as described above.

SELinux also includes some server programs that have been modified to enforce SELinux policies. An example server is the SELinux X server [325]. The X server provides mechanisms that could enable one client to obtain secret information or compromise the integrity of another. This has long been known as a problem [85], and several implementations of access enforcement for windowing systems have been developed over the years [90, 86, 42, 199, 289, 95]. The SELinux community built a reference monitor interface for the X server, and defined a user-level policy server that can respond to authorization queries [43, 308]. The policy server design is general in that it can support authorization requests from multiple user-level processes, similarly to the Flask object managers (see Chapter 7). The aim is that the user-level policy could be verified to ensure that such trusted servers enforce a policy that is compliant with the SELinux system policy.

The SELinux MLS policy contains over 30 subject types that are trusted by the system. In many cases, the subject types are associated one-to-one with programs, but some subject types, such as `init`, have many scripts that are run under a single trusted type. The larger the amount of trusted code, the more difficult it is to verify tamperproofing and verify correctness.

We note here that there are certain programs that SELinux does not trust. For example, SELinux does not trust NFS [267] to return a file securely. As a result, SELinux associates an `nfs_t` type label for all these files, regardless of their label on the NFS server. The reason for this is that the NFS server delivers files to its clients in the clear, so an attacker may reply with a false file upon an NFS request. File systems with integrity-protected communication, such as kerberized Andrew File System [225, 234], could potentially be trusted to deliver labeled files. A variety of distributed files systems that provide secure access to files have been designed [28, 2, 197, 104, 255, 314]. A detailed treatment of this subject is beyond the scope of this book.

9.2.7   SELINUX SECURITY EVALUATION

We now assess whether SELinux satisfies the secure operating system requirements of Chapter 2. SELinux provides a framework in which these requirements can be satisfied (i.e., it is "secureable" like Multics), but the complexity of UNIX-based systems makes it difficult to provide complete assurance that these requirements have been met. Further, the practical requirements of UNIX systems (i.e., the function required) limits our ability to configure a system that would satisfy these requirements. As a result, SELinux provides significant security improvements over traditional UNIX systems (see Chapter 4), but it is difficult to quantify these improvements to the extent required of a secure operating system.

1. **Complete Mediation**: How does the reference monitor interface ensure that all security-sensitive operations are mediated without creating security problems, such as TOCTTOU?

   The Linux Security Modules framework's reference monitor interface is designed to authorize access to the actual objects used by the kernel in security-sensitive operations to prevent vulnerabilities, such as TOCTTOU.

2. **Complete Mediation**: Does the reference monitor interface mediate security-sensitive operations on all system resources?

   The Linux Security Modules framework mediates operations identified by the LSM community to lead to security-sensitive operations. The mediation provided is effectively a union of all the Linux reference monitor prototype's constructed.

3. **Complete Mediation**: How do we verify that the reference monitor interface provides complete mediation?

   Since the LSM framework's reference monitor interface was designed in an informal manner, verification that it provides complete mediation is necessary. Source code analysis tools were developed to verify that the security-sensitive kernel data structures were mediated [351] in a consistent manner [149], and bugs in the LSM reference monitor interface were found and fixed. However, these tools are an approximation of the complete mediation requirements, and they are not applied on a regular basis. Nonetheless, no errors in the reference monitor interface placement have been found since its introduction in Linux 2.6.

4. **Tamperproof**: How does the system protect the reference monitor, including its protection system, from modification?

   The LSM reference monitors, such as SELinux, are run in the supervisor protection ring, so they are as protected as the kernel. Although the LSM framework is a module interface, LSMs are compiled into the kernel, so they can be active at boot time.

   The Linux kernel can be accessed by system calls, special file systems, and device files, so access to these mechanisms must ensure tamper-protection. While Linux system call processing does not provide input filtering at the level of Multics *gates*, work has been done to verify kernel input handling using source code analysis tools [154]. Further, Linux systems provide a variety of other operations that enable access to kernel memory. For example, special file systems, such as the /proc filesystem and *sysfs* filesystem, and device files enable access to kernel memory through files. The SELinux protection state is configured to limit access to trusted processes (i.e., those with trusted subject type labels).

5. **Tamperproof**: Does the system's protection system protect the trusted computing base programs?

   As described above, SELinux system tamper-protection also requires that its trusted user-level programs be tamper-protected. An evaluation of SELinux policy showed that a set of trusted processes which defined a tamper-protected, trusted computing base could be identified [150]. However, several of these trusted processes must be trusted to protect themselves from some low integrity inputs, so satisfying a classical information flow integrity where no low integrity inputs are received (e.g., Biba integrity protection [27]) appears impractical. Some trusted SELinux services (e.g., sshd and vsftpd) were shown to enforce a weaker version of Clark-Wilson integrity [54, 285].

6. **Verifiable**: What is basis for the correctness of the system's trusted computing base?

   As is typical, verifying the correctness of security enforcement is the most difficult task to achieve. Verifying correctness of the implementation of the Linux kernel and trusted programs is a very complex task. For this large a code base, written mostly in nontype safe languages, by a variety of developers, verification cannot be complete in practice. Linux has been assured at the Common Criteria evaluation level EAL4 (see Chapter 12), which requires documentation of the low-level design of the kernel. Converting this low-level design into a model in which security properties can be verified would be a challenging task, and it may be impractical to verify how the source code implements the design correctly.

7. **Verifiable**: Does the protection system enforce the system's security goals?

   SELinux policies define a precise, mandatory specification of the allowed operations in the system. As a result, it is possible to build an information flow representation from the SELinux policies [150] (mentioned above), even one that includes the transition state. Also, the MLS policy ensures information flow secrecy satisfies the simple-security and ⋆-security properties. However, the integrity analysis and MLS policy reveal a significant number of trusted subject types (over 30 for integrity and over 30 for MLS, and only some overlap). Thus, the SELinux approach enables the system to be "secureable," but system developers will need to manage the use of trusted code carefully to ensure the verified security goals are really met.

9.3   SUMMARY

The LSM/SELinux system implements a reference monitor in the Linux operating system. The LSM community emerged from a variety of prototype efforts to add a reference monitor to Linux, and developed a reference monitor interface that was acceptable to the security community (for the most part) and to the mainstream Linux community. The SELinux and AppArmor LSMs have been adopted by the major Linux distributors and are supported by many other distributors. While the resulting LSM framework has been only semi-formally tested, it has generally been a successful addition to the kernel. However, the combination of the Linux kernel and LSM framework is too complex for a complete formal verification that would be required to prove complete mediation and tamperproofing.

The challenge has been how to use the LSM reference monitor interface to enforce security goals. We examined the SELinux system which provides a comprehensive set of services for implementing security policies and a fine-grained and flexible protection system for precise control of all processes. The SELinux approaches demonstrates the complexity of UNIX systems and the difficulty in enforcing comprehensive security. The outstanding challenge is the definition and verification of desirable security goals in these low-level policies. The AppArmor LSM uses a targeted policy to protect the system from network malice, but proof of security goal enforcement will also have to verify requirements, such as information flow.

CHAPTER 10

Secure Capability Systems

A *capability system* [181] is an operating system that represents its access control policy from the subjects' perspectives. Recall from Chapter 2 that Lampson's access matrix [176] identified two views of an access control policy: (1) an object-centric view, called *access control lists*, where the policy is defined in terms of which subjects can access a particular object (the columns of the matrix) and (2) a subject-centric view, called *capabilities*, where the policy is defined in terms of which objects can be accessed by a particular subject (the rows in the matrix).

Although the access control decisions made by capability and access control list systems are the same, the capability perspective provides some opportunities to build more secure systems, but this perspective also introduces some challenges that must be overcome to ensure enforcement of security goals. In this chapter, we identify these opportunities and challenges, and describe capability system designs that can leverage the opportunities while mitigating the challenges.

10.1 CAPABILITY SYSTEM FUNDAMENTALS

A *capability* is a reference to an object and a set of operations that the capability entitles the holder, first formalized by Dennis and van Horn [72]. Such capability references are extended memory references in that they not only provide location or naming information, but they may also provide access rights for that reference [335]. This form of addressing is known as *capability-based addressing* [89]. Thus, a capability is like a house key [128] in that it permits the holder the access associated with the key. When a process needs to access an object, it presents the appropriate capability to the system, much like one would select the appropriate key to unlock a door. If the capability includes the requested operations, then the access is permitted.

An important difference between capability and access control list systems is the process's ability to *name* objects. In an access control list system, a process can name any object in the system as the target of an operation. Typically, the name space of objects can be searched for a specific object (e.g., file names in a directory hierarchy). Then, access is determined by checking the access control list to determine if the subject associated with the process can perform the requested operations. In a capability system, processes can only name objects for which they have a capability. That is, the process can only reference objects to which they have some access (via a capability), then the system can determine whether the capability authorizes the requested operations on that object.

In his book, Levy compares the access control afforded using the real world analogy of a safe deposit box [181]. Two alternatives that the bank may use to control access to your safe deposit are: (1) keep a list of those persons who are authorized to access the box or (2) provide a set of keys that you can distribute to those persons you wish to give access. In the first approach, the bank uses an access control list to authorize each person's request to access your box. In the second approach, the bank provides a capability (i.e., the key) that you may use to create copies to distribute to others to which you may give access. The differences are between the two approaches are mainly in the cost of authorization and the ease of revocation. The capability approach results in less work for the bank to perform authorization, as simply possession of the necessary key is sufficient. However, if you want to remove access for one person, the access control list enables immediate revocation, whereas the capability approach requires the retrieval of keys.

A fundamental requirement of a capability system is that processes not be able to modify or forge capabilities. If capabilities could be forged, then a process could build a capability for any access it desires. To protect capabilities, capability systems traditionally store them in memory protected from the process. For example, the kernel stores one or more *capability lists* or C-lists for each process. When a process wants to invoke an operation using a particular capability, it references this capability in the invocation. There are a variety of ways that capability systems have enabled processes to reference protected memory securely in such invocations. The Plessey System 250 [63] and CAP systems [336] stored process capabilities in special capability segments. Only the kernel could write to these segments, but typical processes could read from their capability segments to invoke their authorized operations. The IBM System/38 uses tagged memory where tag bits are included with every memory word, enabling the system to distinguish capability and noncapability memory [26, 138]. Amoeba and others [216, 12] use password-protected capabilities, where each capability is encrypted using a key only available to the Amoeba kernel.

10.2 CAPABILITY SECURITY

Capability systems have a couple of conceptual security advantages over ordinary systems: (1) they can be used to define process permissions more precisely, enabling *least privilege* [265], and (2) they enable permissions to be more easily transferred from one process to another, enabling the definition of *protected subsystems* [265]. First, capability systems can assign a distinct set of capabilities to each process, so it is possible to assign only the capabilities required for each process based on its specific purpose. This contrasts with an access control list system where all the permissions for any possible use of a program must be assigned. In a capability system, we can customize a process's capabilities based on the particular execution of that process's program to minimize the set of permissions while still providing the necessary function.

Second, capability systems enable processes to copy their capabilities for other processes (again, like a house key), so a protected subsystem may perform operations on behalf of clients without the need to be assigned the permissions of all clients. For example, a subsystem that provides access to a database does not need to have its own permissions to the database entries. When it is invoked, the client will provide those capabilities with the request, so that the subsystem need not accidentally give the client unauthorized access.

This problem is known as the *confused deputy problem* [129], due to Hardy, and it is inherent to ordinary operating systems. In an ordinary operating system, permissions are assigned to objects and it is inconvenient to change which subjects can access an object at runtime. Thus, the subsystem would be assigned the rights to operate on behalf of any client (i.e., subsystems have the union of all their clients' rights to the objects that they serve). As a result, if the subsystem is confused, it could provide unauthorized access to another client's data. Whereas in a capability system, this would not be possible because the subsystem would only be able to perform operations using the capabilities that the client could provide.

Many in the operating systems community saw capability systems as the proper architecture for constructing fail-safe operating systems. The capability abstraction was first defined by Dennis and Van Horn [72], and was quickly adopted in a variety of systems [344, 178, 4]. Initially, capabilities were implemented in software, but performance concerns led to the development of several systems that implemented capabilities in hardware [63, 336, 26, 238]. In the 1970's, it was envisioned that capability systems may provide better flexibility and security than ordinary operating systems, such that they would supplant ordinary operating systems [186]. In more recent years, capabilities have been used to support single address space operating systems [133, 45] and distributed operating systems [216, 65].

10.3   CHALLENGES IN SECURE CAPABILITY SYSTEMS

The fundamental question is whether a capability system architecture is suitable for implementing a secure operating system. Definition 2.5 requires that a secure operating system meet mediation, tamperproofing, and verifiability requirements when enforcing a mandatory protection system. While the requirements for mediation and tamperproofing can be met, the flexible distribution of capabilities fundamental to capability systems present some problems in verifying the enforcement of security goals (i.e., the protection system is not mandatory). In this section, we identify these problems, and we describe how capability systems have been modified to address these problems in the following section.

First, we describe how capability systems satisfy the secure operating system requirements of mediation and tamperproofing. Since capabilities must be used to name objects, access control is inherently bound to access operations in capability systems. For example, in order to access a file, a process must provide a capability that both identifies which file to access (i.e., naming the file as described above) and specifies the rights that the process has over that file. Without the capability, the file cannot even be found, so mediation is fundamental to capability systems.

Tamperproofing requires that the trusted computing base, including the reference monitor and protection state itself, cannot be modified by untrusted processes. Like an ordinary system, a capability system provides untrusted processes with access to services offered by trusted processes, such as the kernel and protected subsystems. For example, if an untrusted process has a capability that enables the execution of a trusted process, the trusted process must protect itself from any input the untrusted process may provide. These problems are not significantly different than for ordinary operating systems, as they may accept requests from untrusted processes (e.g., network communications) or be executed by an untrusted process (e.g., via `setuid`). Capability systems are also fundamentally aware of the need to prevent unauthorized modification of the protection state. As discussed above, the prevention of capability forgery are design principles of such systems.

The flexibility that processes have to distribute capabilities in capability systems conflicts with the desire to prove that particular security goals are truly enforced, so ensuring that security goals are verifiable is problematic in capability systems. Researchers have identified three specific problems in ensuring the enforcement of security goals in capability systems: (1) care must be taken to ensure that the $\star$-property in the Bell-LaPadula policy is correctly enforced [23]; (2) capability systems require addition mechanism to ensure that each protection state is *safe* [130], enforcing the system security goals; and (3) changes in security requirements (i.e., policy) require that capability systems *revoke* all newly unauthorized capabilities. In general, these problems are created by the discretionary distribution of capabilities inherent to the model, so solutions focus on how to add mandatory boundaries that ensure operation that satisfies system security goals.

10.3.1   CAPABILITIES AND THE $\star$-PROPERTY

Boebert [32] and Karger [158] note that traditional capability systems fail to implement the $\star$-security property of the Bell-LaPadula policy [23], as shown in Figure 10.1. Suppose there are two processes, a high secrecy process A that has access to high secrecy data and low secrecy process B that is not authorized to access that data. If our goal is to implement a multilevel security (MLS) policy such as Bell-LaPadula, then the high secrecy process may read data in the low secrecy process. For example, A uses its legal capability *B1 read* to read segment $B1$. Since capabilities are data, the high secrecy process A can read the capabilities (e.g., *B2 write* in the low secrecy process's segment $B1$. Then, high secrecy process A has a capability to write its secrets (e.g., data from segment $A1$) to a low secrecy segment $B2$, violating the $\star$-security property.

While it may be unlikely that an error in a high secrecy process may result in such a leak, remember that secure operating systems must prevent any code running in a high secrecy process, including malware, such as Trojan horses, from leaking data. A Trojan horse could be designed that retrieves write capabilities to low secrecy files to enable the leak.

10.3.2   CAPABILITIES AND CONFINEMENT

Karger states that the violation of the $\star$-property implies that capability systems fail to enforce process *confinement* [158]. Lampson defined confinement in terms of [177]: (1) processes only being able to communicate using authorized channels and (2) process changes not being observable to unauthorized processes. The failure above in implementing the $\star$-property does result in an unauthorized communication channel, but the problem is even broader than this: we must ensure that no unauthorized communication is present for any security policy.

Consider a second example from Karger [157]. An attacker may control a program P. When an unsuspecting victim provides a capability C to P, the malicious program can store the capability. This enables the attacker to use this capability, presuming that the attacker can run at the same

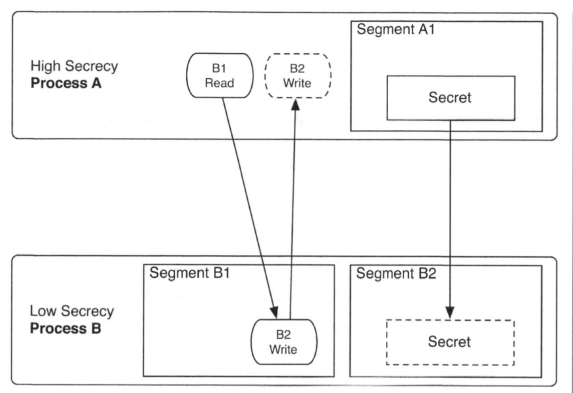

Figure 10.1: A problem with the enforcing the ⋆-property in capability systems

clearance as the victim. Clearly, the attacker should not have access to this capability, but how does the kernel know that a capability stored by a program cannot be used in another context?

Confinement is not achieved in the example above because the program P has the discretion to give the capability to the attacker. Like other mandatory access control systems, we want to define a mandatory policy that ensures that the program P cannot give the victim's rights away.

10.3.3 CAPABILITIES AND POLICY CHANGES

The third problem is the *revocation problem*; capabilities are difficult to revoke. Recall Levy's safe-deposit box example. When keys are distributed among the authorized people, the owner of the safe-deposit and the bank lose the ability to restrict who has access. Should the owner or bank try to change who can access the safe-deposit box after the keys have been distributed they have a couple of challenges in enforcing this change. First, they have to locate all of the keys that were given out. While they may know how many keys were created and to whom they were initially distributed, the keys may not longer be in the possession of those people and they may not remember what they did with them. Second, keys may have been copied, such that it may not be possible to determine whether all the keys have been revoked. In general, it may be easier to change the lock and start again.

This analogy means that should we decide to change the security goals that we want to enforce in a capability system, we may not be able to determine whether we have accounted for all the capabilities necessary to prove that the new goal can be achieved. We not be able to find all of the capabilities to all objects or some unknown copies may have been made. Mechanisms to bring the set of capabilities back into some approved state may be expensive (search all memory) or disruptive (delete all capabilities and start again).

10.4   BUILDING SECURE CAPABILITY SYSTEMS

Work in secure capability systems aims to address these problems to enable effective verification that the system enforces a well-defined set of security goals. While a variety of capability system designs have been modified to solve these problems, we focus on two capability systems: SCAP [157] and EROS [286]. Both SCAP and EROS are capability system designs based on existing designs, CAP [135, 223] and KeyKOS [128], respectively, but extended to solve these fundamental problems. Table 10.1 shows a summary of how EROS and SCAP address the three problems in capability systems. We develop and compare these solutions below.

Table 10.1: Summary of SCAP and EROS solutions to the major security issues in capability systems.

Security Issue	SCAP Solution	EROS Solution
*-Property	Convert to *read-only* capabilities by MLS policy	Define *weak* capabilities that transitively fetch only read-only capabilities
Confinement	Use Access Control List to define confinement	Define *safe* environments for confined processes or test via *authorize* capabilities
Revocation	Revocation by eventcounts (single page entry) or revocation by chaining (multiple page entries)	Indirect capabilities that permit later revocation of all descendants (similar to Redell [252])

10.4.1   ENFORCING THE *-PROPERTY

The SCAP design to ensure that the *-property is not violated in capability systems leverages two key insights [157]: (1) capabilities must be loaded into a capability cache prior to use and (2) we simply need to remove unauthorized access from any capability loaded into the cache to prevent leakage. SCAP requires that a process must load a capability into its capability cache (i.e., its capability list or C-list) prior to using it. This load operation provides the operating system with a point of complete mediation to inspect the capabilities being loaded. This mediation can be used to determine whether the capability provides write access to an object with a lower access class than the process (i.e., where write permission would violate the Bell-LaPadula policy). Enforcement of such an approach requires the SCAP kernel to include labels with capabilities, labels with processes, and an MLS access policy, so that the kernel can assess whether the capability may be loaded legally.

Other capability systems, the Secure Ada Target (SAT) [34] and the Monash capability system [12] implement similar enforcement semantics, albeit with markedly different approaches. SAT implements semantically similar checks as SCAP to ensure that any capability being loaded adheres to an MLS policy, but the enforcement is done entirely in hardware. Monash's solution is also semantically similar, but Monash is a password capability system which limits access to write capabilities by keeping the encryption keys used for these capabilities secret. Only authorized processes can obtain the key from the system.

Rather than just providing a point of mediation to decide whether to reduce capability permissions, the EROS system defines a capability that automatically generates the correct permissions [287]. EROS defines the notion of a *weak* attribute for a capability, the combination of which we will call a *weak capability*. If a weak capability is used to fetch some other capabilities, all the retrieved capabilities are automatically reduced to read-only and weak capabilities. Like SCAP, the reduction is performed when the capabilities are loaded into the capability cache. Unlike SCAP, EROS enforces the ★-property without the need to consult an MLS policy at runtime, a potentially significant performance advantage [288].

Instead, EROS requires that an MLS policy be consulted when capabilities are created (e.g., at load time) to determine when a process should be given a weak capability. In order to ensure that this works, the system must ensure that any capabilities constructed that enable a process to access lower-secrecy memory must be weak capabilities. As EROS does not define how capabilities are initially distributed [1], a higher-level service is necessary that understands the labeling of processes and capabilities, and has access to a Bell-LaPadula policy [23], so the ★-property is enforced correctly.

The idea of a weak attribute is based on *sense* capabilities in KeyKOS [128] which only enable retrieval of read-only capabilities (i.e., in most cases), even if the capability fetched has read-write privileges. The EROS design generalizes this by making the semantics uniform and transitive. First, weak capabilities have the same effect regardless of the type of the object referenced by the capability. Second, if a sequence of capabilities is required to retrieve some target capability, the target capability is reduced to read-only and weak if any capability in the retrieval sequence is weak.

10.4.2 ENFORCING CONFINEMENT

The confinement problem was identified by Lampson in 1970 [177], so several capability system designs in the 1970s aimed to provide confinement guarantees. The HYDRA capability system [56, 343] provided confinement by defining *confined protection domains* that were not allowed to store

---

[1] EROS does check that the capabilities available to a process being to an authorized set, which also is defined external to EROS, as described below in Section 10.4.2.

capabilities or regular data into potentially shared objects. This prevents leakage of secret data, but does not permit sharing that is legal under Bell-LaPadula model.

The Provably Secure Operating System (PSOS) [92, 226] is a design for a capability system that also provides an approach for confinement [2]. In PSOS, *secure documents* are defined such that write capabilities may not be stored in such documents. This prevents propagation of write capabilities that would enable leakage. The rules for checking whether a write capability may be stored are complex, as they pre-date the definition of the Bell-LaPadula model.

In practice, there are two ways to enforce confinement: (1) build an execution environment for the process that satisfies confinement requirements or (2) verify that confinement is preserved whenever an access right is obtained. Both HYDRA and PSOS take the former approach. They define restrictions on execution environments that enforce confinement requirements. In the second approach, the system ensures that whenever a process obtains a capability (or access right in general) the confinement requirements of the system are met, or the capability is revoked.

While the first approach is conceptually cleaner and more efficient to execute, it has a fundamental limitation. This limitation is the intractability of the *safety problem* [130]. A system is said to be *safe* if all future protection states in a protection system only grant authorized permissions to processes (i.e., confine the processes correctly). Harrison, Ruzzo, and Ullman showed that determining whether an arbitrary protection system prevents a principal from obtaining an unauthorized permission is undecidable. That is, given a current protection state and the operations in a general protection system, we cannot determine whether there may be some future protection state in this system which some principal may obtain unauthorized access.

In addition to HYDRA and PSOS, general access control models have been defined in which verifying safety is tractable, such as the Take-Grant model [155, 31], Typed Access Matrix [270], and Schematic Protection Model [269]. However, all these models require limitations in the possible policy designs that have proven unacceptable in practice.

Unlike these systems, SCAP enforces confinement by mediating each change in protection state. In SCAP, changes in protection state require the loading of capabilities into SCAP's capability cache, so this serves as an effective point to enforce confinement as well. As described above, SCAP verifies that a process can load a capability into the cache by verifying that the process label permits access to write objects with capability's label using an external policy. To enforce the $\star$-property, an MLS policy is checked, but any confinement policy may be used in general. Interestingly, SCAP uses an access control list to define confinement policies, resulting in SCAP enforcement being based on a combination of capabilities and access control lists.

EROS aims to provide confinement with a pure capability approach, rather than the hybrid approach of SCAP [287]. Also, unlike SCAP, EROS provides confinement through the creation of safe execution environments. EROS defines a *safe* execution environment as one that contains only safe capabilities. A *safe capability* meets the following requirements:

1. It trivially conveys no mutate authority, or

---

[2]PSOS was the basis for the Secure Ada Target [34] mentioned above.

2. It is a read-only, weak capability, or

3. It is a capability to a constructor that (recursively) generates confined products (i.e., environments).

These capabilities preserve confinement because they do not change the protection state by adding rights (#1 and #2) or they only add rights by generated new confined environments.

These requirements for a confined environment are very restrictive, so EROS includes a fall-back position that is similar to that of SCAP. If a process has some capabilities that are not safe, EROS checks whether these capabilities are *authorized*. That is, it checks whether the capabilities that the confined environment may ultimately obtain are a subset of the authorized capabilities that define confinement. If so, the process may be executed. Determining all the capabilities that a process may need in advance is nontrivial undertaking, so in some cases, we envision that runtime checking similar to SCAP, but using authorized capabilities to define confinement would be necessary. Since EROS also mediates capability loads to enforce weak capabilities, this would be feasible.

The SCAP design also considers the impact of covert channels on confinement carefully. SCAP prevents storage channels by eliminating system-wide views of system state. For example, domains are limited to storage quotas and page tables are unshared. Addressing timing channels is a more ad hoc procedure, but is a consideration throughout the design.

10.4.3 REVOKING CAPABILITIES
Redell defined the first comprehensive approach to revocation in capability systems [252]. In this system, the owner of an object has a choice whether to grant normal capabilities (i.e., with no hope of revocation) or grant capabilities that are associated with a special *revoker capability*. A revoker capability is a level of indirection that enables the owner to revoke all the capabilities that reference the object through that revoker capability. Simply by deleting the revoker capability, the other capabilities lose access to the object. This is because the object is only available via the revoker capability.

The idea is shown in Figure 10.2. An owner creates a revoker capability and grants a capability that points to that revoker capability to Process 1. Revoker capabilities may be used by other subjects as well. Process 1 can also create a revoker capability and create a capability that points to it for Process 2. Thus, the owner will revoke both Process 1 and Process 2's access should she delete her revoker capability. However, Process 1 can only revoke access from Process 2. Note that the capability for Process 3 cannot be revoked (i.e., without examining Process 3's memory) because it points directly to the object.

Redell's scheme could result in a deep nesting of revoker capabilities, so SCAP defines two different schemes, called *revocation with eventcounts* and *revocation by chaining*. Revocation with eventcounts is appropriate for systems that use the same page table for each shared object. In revocation by eventcounts, an event, such as revoking a capability, causes an eventcount to change. Eventcount values are stored with capabilities as well, so that should a revocation occur, the event-

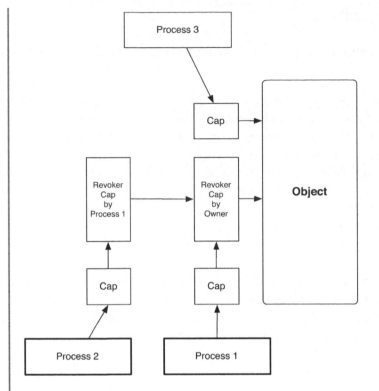

Figure 10.2: Redell's revoker capability approach: When the revoker capability is revoked all the capabilities that were based on it are also revoked.

count between the capability and the page table will differ, triggering a verification whether the capability is still valid.

If there are multiple page table entries that point to the same physical page (because it is shared by multiple processes), the revocation by eventcounts cannot be used. Revocation by chaining creates a ring of capability records for the same page by adding a pointer field to each capability. Thus, the revocation of any capability in the chain enables triggers a reassessment of the validity of the remaining capabilities in the chain. All such capabilities are accessible because they are chained together.

Both the revocation by eventcounts and revocation by chaining approaches are rather complex and potentially expensive to implement, so the later EROS system reverted to an indirection mechanism similar to Redell [252] to revoke capabilities. An indirect (revoker) capability may be obtained that enables later revocation, as described above [288]. The memory usage problems cited by SCAP as a reason for seeking alternative revocation schemes had become less of an issue by the late 1990s.

10.5   SUMMARY

In the chapter, we examine the construction of secure operating systems from capability systems. Capability systems have conceptual advantages in enforcing security because they can be used to define protection domains specific to a particular execution of a program easily and they enable permissions to be distributed with program invocation preventing the *confused deputy problem* [129] by limiting the user of others' permissions. However, capability systems also have sme inherent security problems brought about by the discretionary nature of capability management.

The SCAP and EROS capability systems address these limitations by adding mandatory restrictions on the use of capabilities to ensure *safe* system behavior. They each define mechanisms to limit the capabilities that a process can receive to only those within the system's security goals (e.g., *weak capabilities* of EROS), but restricting the system's behavior in a sufficiently flexible manner requires runtime checks (e.g., on capability loading). Revocation is a conceptual problem for capability systems, but in practice simple ideas, such as Redell's indirect *revoker* capabilities, appear sufficient. Thus, the biggest challenge for capability systems, like many systems, is providing a practical execution environment that can be proven to ensure system security goals.

CHAPTER 11

Secure Virtual Machine Systems

A problem in building a new, secure operating system is that existing applications may not run on the new system. Operating systems define an application programmer interface (API) consisting of a set of system calls it supports. New operating systems often define new APIs, resulting in the need to port applications and/or write new applications. This has been a major problem in gaining acceptance for secure operating systems based on emerging kernels, such as Trusted Mach [35] and Flask [295], and the reason for so much focus on securing commercial systems.

An alternative that enables execution of multiple operating systems on one computer is called a *virtual machine system*. A virtual machine system enables multiple operating system instances and their applications to run concurrently on a single physical machine. Each operating system instance runs in a virtualized environment that emulates a physical platform, called a *virtual machine* (VM). While each operating system still manages the state of its applications and abstractions (e.g., files, sockets, processes, users, etc.), it does not actually manage use of the physical hardware.

In virtual machine systems, a new component is introduced, called a *virtual machine monitor* (VMM), that multiplexes physical system resources among the operating systems in the VMs. The virtual machine operating systems no longer control the use of system hardware, but instead receive hardware access only via an indirection through the VMM. Only the VMM runs in supervisor mode, so only it has access to system hardware directly.

VMM architectures are classified into two types, as shown in Figure 11.1, distinguished by whether the VMM runs directly on the hardware (Type (1) or on another *host* operating system (Type 2). Since a Type 2 VMM requires the services of a host operating system in order to run, its trust model must include this operating system (and perhaps it is insecure). As a result, the trusted computing base of a Type 1 VMM can be smaller than that of a Type 2 VMM. Examples of Type 1 VMMs include IBM's VM/370 [69], Xen [346], and VMWare's ESX Server [319]. Examples of Type 2 VMMs include VMWare's GSX Server [320], Microsoft's Virtual PC [208], and User-Mode Linux (UML) [77]. Of course, Type 1 VMMs must also provide system services, such as IP networking, so care must be taken to keep the Type 1 VMM's trusted computing base as small as possible.

There are variants Type 2 VMMs as well where all the VMs use the same kernel interface, so all VMs use the host operating system for handling all system calls. Such systems include Solaris Containers (Zones) [175] (see Chapter 8) and FreeBSD Jails [156] (see Section 7.5.3). In both of these systems, the host operating system creates isolated computing environments akin to virtual machines, but all the system calls from the virtualized systems are forwarded to the host operating system. There are no guest operating systems in these systems.

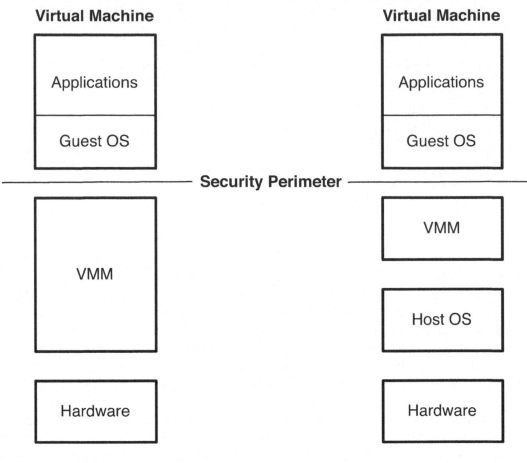

Figure 11.1: Type 1 and Type 2 VMM Architectures: A Type 1 architecture runs directly on the hardware, whereas a Type 2 architecture depends on a host operating system.

Virtual machine systems were originally envisioned as a means to provide better resource utilization, by enabling multiple software systems to use a single system's hardware. Virtual machine system designs sometimes focus on supporting legacy code [39, 69], but other virtual machine designs have focused on security, or at least isolation, as a main goal [161, 259, 332]. In general, virtual machine system designers now expect to: (1) support legacy applications effectively, with little or no modification; (2) provide effective isolation and security enforcement; and (3) with a modest performance overhead that would be undetectable given modern hardware. For security, virtual machine systems have a couple of significant advantages over traditional systems.

First, virtual machine systems offer the potential of reducing the size of the trusted computing base. Since operating systems run in VMs, it is possible to remove the operating system from the trusted computing base. The VMM now becomes the trusted computing base. However, whether this *actually* reduces the size of the system's trusted computing base depends on how the VMM is designed. While a few have specially-designed, small code bases, many VMMs depend themselves on a complete, ordinary operating system (even Type I VMMs).

Second, virtual machine systems provide an additional, coarser-grained layer of control for securing a system. The hope is that coarser-grained enforcement translates into simpler mediation and simpler policies. With respect to the reference monitor guarantees, virtual machine systems should make it easier to ensure complete mediation and make it easier to verify that security goals are met. First, a VMM reference monitor only needs to mediate the distribution of resources among VMs and inter-VM communication. Since resources are typically partitioned among VMs (i.e., they are not shared), securing resource distribution is often simpler. Once a VM is started, only communication with other VMs needs to be mediated. Typically, only a small number of primitives enable inter-VM communication. Second, the VMM policy will likely be simpler than a secure operating system policy. Because the number of VMs is smaller than the number of processes running on an operating system and the number of possible inter-VM communications is smaller than the number of resources in a traditional system (e.g., files), it should be easier to verify that the VMM policy expresses the intended security goals.

Rushby identified the reasons above as motivation for moving from the security kernel approach (see Chapter 6) to a new approach, called a *separation kernel* [259, 260]. We first review Rushby's arguments for the separation kernel architecture. We then examine the design and implementation of a secure virtual machine system, the VAX VMM Security Kernel [161]. We then examine issues in building secure operating systems in the context of the current virtual machine systems.

While the additional layer of control offered by virtual machine systems presents these potential benefits, building secure virtual machine systems is not without challenges [107]. First, virtual machine systems may generate problems for administrators, as there will be many more VMs to administer and disinfect than physical machines. Second, the ability to save, restore, and migrate VMs may generate difficulties in ensuring that the VM software base is current and consistent with organizational requirements. Third, the identity of virtual machines will be harder to determine than the identity of physical machines. Virtual machines may even be able to migrate across administrative domains. Fourth, data leaks of VMs into various physical systems and the integrity impact of individual systems into VMs may be difficult to track [51, 52]. Virtual machine security solutions must account for these challenges in order to leverage its benefits.

11.1 SEPARATION KERNELS

In 1981, Rushby examined the difficulty of building security kernel systems, such as Scomp and GEMSOS discussed in Chapter 6. Rushby found that security kernel systems, despite their near-minimal trusted computing base, had a significant, uncontrolled reliance on trusted services. As a result, he defined an alternative approach that he called a *separation kernel* [259, 260].

In a multilevel (MLS) system, if any process can write data from a higher sensitivity level to a lower sensitivity level, it violates the MLS policy (see the Bell-LaPadula policy [23] in Chapter 5). However, some services are entrusted with such operations, such as inline encryption systems [259], that encrypt secret data and send it via public networks. Also, other services may be trusted to process data at multiple sensitivity levels with leakage, such as file and print servers. In an MLS system, such processes are simply trusted, and the MLS policy is not enforced on them.

Rushby claimed that ensuring the correct behavior of trusted services of a security kernel system is too complex. In a general purpose system, we have a large number of trusted services, potentially complex interactions among trusted services, and a variety of interfaces accessible to untrusted processes. The SELinux system with its 30+ trusted programs is indicative of the number of trusted programs in a general-purpose system. The interactions among the resulting trusted processes are not clearly identified, but they are likely to be complex. Most dangerous of all is the number of ways that untrusted processes may invoke trusted programs. In minimal security kernel system, such as Scomp and GEMSOS, there were 30–40 gates defined to control such invocations. In a modern operating system, there are hundreds of system calls.

Rushby's solution is to treat each trusted program as one would a single node in a distributed system. For example, a file server node would be a single-purpose system attached to other systems via a single communication channel. If the "file server adheres to and enforces the multilevel security policy, the security of the rest of the system follows" [259]. That is, enforcement of system security goals can be composed from isolated elements that "adhere to and enforce" [259] those security goals.

Rushby coined the name of such a system as a *separation kernel* to distinguish it from a security kernel. A separation kernel emphasizes independence and authorized communication. Each trusted service runs in a isolated and independent system, perhaps on the same physical platform or perhaps not, and the services can only be accessed by a small number of mediated communication channels. The separation kernel is capable of complete mediation of such communication channels, such that the services could be isolated completely from the remainder of the system.

Rushby noted the similarity between the separation kernel concept and virtual machine systems, as we described them above. The major distinction is that separation kernels do not require that the separation kernel provide a virtualized hardware API, as a virtual machine monitor does. The trusted services in a separation kernel may be customized to a separation kernel system and minimized (e.g., not run a guest operating system). With increasing popularity of paravirtualized hypervisors, such as Xen [19], which require some awareness of running on a virtual machine monitor, and custom VMs as proposed for Terra [105], the line between a separation kernel and a virtual machine system is becoming blurrier each year.

A particular family of systems that implement the separation kernel approach are called *Multiple Independent Levels of Security* (MILS) systems [131, 7, 193, 9]. A MILS system architecture is shown in Figure 11.2. A each service runs in an isolated *regime* supported by a MILS middleware

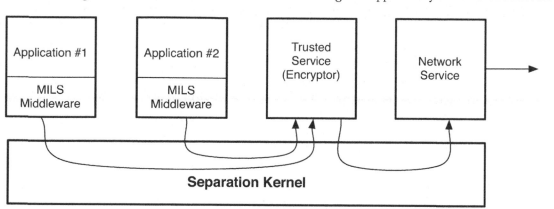

Figure 11.2: A *Multiple Independent Levels of Security* (MILS) system architecture: individual systems are treated like separation physical machines and unsafe operations (e.g., encryption of secret data prior to network delivery) are routed through a simplified and verifiable trusted service.

layer. Coarse-grained communication channels, analogous to network communication are provided, and an MILS separation kernel mediates access. If a trusted service is required (i.e., is trusted to enforce MLS as described above), it placed in its own regime, and the MILS middleware forwards unsafe requests to such services automatically. For example, Figure 11.2 shows a service that encrypts network traffic before being forwarded to the network service. Such trusted services should be small, so they may be verifiable. The MILS separation kernel architecture has been applied to mission-critical deployments for some time, but it is just now starting to garner attention in the mainstream. A proposal for how to construction and evaluate MILS systems has been proposed [233] (i.e., a *protection profile*, see Chapter 12), as has a critique [347].

11.2   VAX VMM SECURITY KERNEL

The VAX VMM Security Kernel is a virtual machine system that aims to achieve the goals of a secure operating system [161]. The VAX VMM runs untrusted VMs in such a manner that it can control all inter-VM communications, even covert channels. The key features of the VAX VMM are its implementation of virtualization, which ensures that the untrusted VMs cannot circumvent the authority of the VMM, and its layered system architecture, which improves security via modularity and minimizing interdependencies.

The VAX VMM design was begun in 1981 from a discussion between Paul Karger and Steve Lipner. The project was motivated by the KVM/370 system, which was a retrofit of the security into the existing IBM VM/370 system [114]. The KVM/370 design was limited by the need to reuse the code from the existing VM/370 system. The KVM/370 design isolates VMs into separate security classes within the architecture of the VM/370 system by adding a layer between the VM/370

VMM and the untrusted VMs, called the Non-Kernel Control Program (NKCP). An NKCP was created for each MLS sensitivity level (e.g., secret and confidential), and they ensured that all VM communication across sensitivity levels was mediated. However, the addition of this additional layer means that access to VMM functions generated *two* context switches, one to NKCP and one to the VMM. This had a significant effect on system performance, reducing the performance by a factor of 2–10 [114]. In addition, the VAX VMM had more extensive support for security management and covert channel prevention not present in the KVM/370 system.

11.2.1 VAX VMM DESIGN

The architecture of the VAX VMM Security Kernel system is shown in Figure 11.3. The VAX VMM

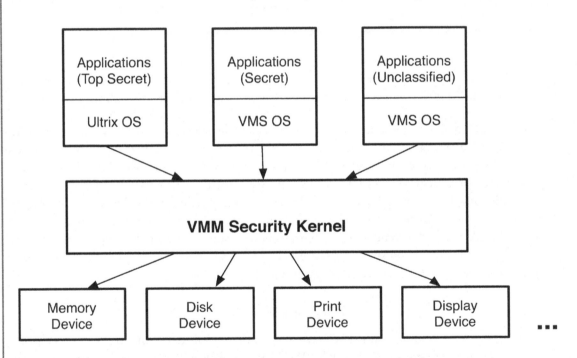

Figure 11.3: The VAX VMM System Architecture: *virtual machines* are labeled according to the secrecy of the information that they can obtain through the *VMM security kernel* to the system's physical devices.

is a Type 1 VMM, in that it runs directly on the hardware. The VMM includes services for storage (on tape and disk) and for printing. In general, the VMM must multiplex all physical devices among VMs, but some key devices, in particular Ethernet networking devices, were not supported, which became problematic for the system's deployment. Each VM could run at its own clearance, and the VMM includes a reference monitor that ensures that any use of physical resources is mediated according to a mandatory access control policy.

The VMM architecture results in coarser-grained physical resources than for a traditional operating system. The VAX VMM views access to individual devices and storage (i.e., disk and tape) volumes as objects. Thus, its reference monitor mediates access at this level. The VAX VMM also provides an abstraction called *virtual disks*, which partitions the physical disk into isolated chunks that may or may not correspond to the physical disk boundaries. The use of virtual disks enables two VMs at different clearances to share a single physical disk securely (e.g., preventing one VM from accessing the other's data).

The VAX VMM enforces both secrecy and integrity requirements upon its VMs. Versions of the Bell-LaPadula secrecy model [23] and Biba integrity model [27] (see Chapter 5) are supported. In order to express permissions that do not fit into these information flow models, the VAX VMM provides the means for specifying additional *privileges*. For example, if a user is permitted to see some data at a higher secrecy level than she is allowed by her clearance, a *user privilege* can be specified to permit such access. The use of such privileges requires a *trusted path* (see Chapter 7), and such uses are audited by the VMM.

The VAX VMM itself is designed in a layered fashion, motivated by the Multics design work of Janson [151], Reed [254], and the Naval Postgraduate School [67]. The idea is that each layer adds well-defined VMM functionality, and no layer depends on functionality provided by a higher (i.e., less-trusted) layer. For example, the I/O layer provides device I/O (i.e., supplies the VMM system drivers), and the VM physical memory layer uses some of these drivers to manage physical memory. The VM virtual memory layer uses both to implement manage access to a prescribed amount of physical memory per VM and ensure proper mapping of device resources (e.g., storage via the I/O layer) to virtual memory.

In building the VAX VMM security kernel, four major challenges had to be addressed: (1) virtualizing the protection rings of the VAX processor; (2) identifying the sensitive instructions of the VAX processor; (3) emulating I/O operations generated by the untrusted user VMs; and (4) enabling the VAX VMM system to be self-virtualizable. The first two challenges ensure that the VMM's reference monitor truly provides complete mediation. The second two challenges provide function in a secure manner.

First, the VAX VMM designers had to add a new virtual ring to the VAX processor. The VAX processor supports four rings: kernel, executive, supervisor, and user. A traditional VAX OS (VMS) ran in the kernel ring, but also some specific VMS system software ran in the executive and supervisor rings as well. Thus, all four physical rings were already used. The VAX VMM design runs the VMM in the kernel ring, and compresses the OS kernel and executive into a single ring. While some protections between kernel and executive code are added, in general, the OS kernel may not be protected from bugs in the executive since they run in the same ring.

Second, even though the OS kernel is run a higher ring than the VMM, the processor may still permit the higher ring to run security-sensitive operations, thus circumventing complete mediation. Processor instructions that may be run only in the privileged ring (i.e., the kernel ring in the VAX architecture) are said to be *privileged*. Processor instructions that reference or modify sensitive data in the privileged ring are said to be *sensitive*, and are security-sensitive in building a VMM reference monitor. Popek and Goldberg [247] state that for an architecture to be virtualizable, all sensitive instructions must be privileged. However, for the VAX instruction set, this was not the case, so the designers had to extend the VAX architecture (i.e., change the VAX microcode actually) to indicate whether code was running in a VM, so that the VMM could emulate its execution securely. For example, the hardware register that stored the current ring number had to be emulated to prevent the OS from discovering that it is being virtualized (e.g., to prevent a false error from being raised).

Third, in addition to emulating sensitive, but unprivileged instructions, the VAX VMM must also emulate access to I/O devices. Since only the VMM runs in kernel, devices are no longer accessible to the VM's guest operating systems (i.e., all physical devices are only accessible from the VMM). Virtualizing I/O access for the VAX processor was especially difficult because I/O was implemented by reading and writing registers that are mapped to physical memory. In the guest VMs, these addresses would no longer mapped to the real registers. To emulate this, the VAX VMM required a small change to kernel software to invoke a ring trap once the I/O memory is updated. The VMM then uses the VMs I/O memory setup to process the real I/O request.

The result of making all sensitive processor instructions privileged is that the VAX VMM system is self-virtualizable. A *self-virtualizable* VMM can run in one of its own VMMs, permitting recursive construction of VM systems. Self-virtualization is a nontrivial property of processors. Until the introduction of the Intel VT and AMD Pacifica processors, no x86 architectures were self-virtualizable, thus any VMM design for x86 (see Section 11.3) had to address similar problems as in the VAX VMM design. Ways in which the x86 is not self-virtualizable are described by Robin and Irvine [258].

11.2.2 VAX VMM EVALUATION

We evaluate the security of VAX VMM system using the reference monitor principles stated in Chapter 2. The VAX VMM design made several considerations to prevent covert communication channels, in addition to the overt channels controlled by the reference monitor interface. However, it is difficult to guarantee reference monitor tamperproofing and verifiability in real systems, as we will show. Nonetheless, the VAX VMM system has been carefully designed with security in mind, and aimed for an A1-assurance according to the Orange Book [304].

1. **Complete Mediation**: How does the reference monitor interface ensure that all security-sensitive operations are mediated without creating security problems, such as TOCTTOU?

   The requirement for mediation in the VAX VMM is that all security-sensitive instructions in the VAX processor's instruction set be privileged. As a result, all the instructions that enable updates of VMM state or enable communication via I/O are trapped to the VMM for mediation. Instruction-level mediation requires provides access to all the system resources in need of modification, so TOCTTOU is not a problem.

2. **Complete Mediation**: Does the reference monitor interface mediate security-sensitive operations on all system resources?

   In the VAX VMM design virtualization enables mediation of VM operations, such as disk volume and device access. Since the VM OS cannot access any privileged instruction and the changes to the VAX microcode for the VAX VMM ensures that all sensitive instructions are privileged, the VMM has the opportunity to mediate all security-sensitive commands.

3. **Complete Mediation**: How do we verify that the reference monitor interface provides complete mediation?

   Privileged instruction traps define the reference monitor interface which provides a reliable mediation if indeed all security-sensitive operations are privileged. Note that the VAX VMM depends on modified operating systems to ensure that I/O commands take the necessary trap. If an attacker can control the VM's operating system code, she can cause a sensitive I/O operation to circumvent the mediation.

4. **Tamperproof**: How does the system protect the reference monitor, including its protection system, from modification?

   The VAX VMM reference monitor and protection system are contained within the VMM itself in ring 0. A small amount of user administration is possible from the VMs based on *SECURE* commands. These require a trusted path between the user and the VAX VMM. How the system distinguishes between trusted users and untrusted users is unclear.

   Also, there are several interfaces by which the untrusted VM code can invoke trusted VMM code, such as the execution of sensitive instructions and I/O emulation (i.e., device access). While the design of the VAX VMM mediates all these entry points, the design must further ensure that all the data received from the untrusted VM is handled properly. Unlike Multics, which uses *gatekeepers* to ensure that malicious input data is filtered, the VAX VMM provides no explicit mechanism for such filtering. We imagine that such filtering is done in an ad hoc manner, but its description is not detailed.

5. **Tamperproof**: Does the system's protection system protect the trusted computing base programs?

   The VAX VMM contains no trusted code outside of ring 0. Administrators use a trusted path to access code running inside the VAX VMM to perform administrative operations. Other trusted services, such as authentication are also performed in the *secure server* inside the VMM. This approach is unique to the systems that we have examined. On the positive side, it limits assurance to the VMM itself. Also, the required use of a trusted path to access administrative services significantly limits the ways that trusted code can access such services. On the negative side, any change in trusted services requires changes to the VMM.

6. **Verifiable**: What is basis for the correctness of the system's trusted computing base?

In order to assure correctness of the VAX VMM software to the A1-assurance level of the Orange Book [304], the design [1] was formally specified and analyzed for verify that it satisfied the security policy model. In addition, the system implementation was informally shown to be consistent with the design. Further, the development process of the VAX VMM was tightly controlled. All design decisions and code were reviewed, including any changes to either. The filtering of untrusted input, mentioned under tamperproofing above, was tested thoroughly for response to a variety of legal, illegal, and malformed requests. Even the CPUs were tested for correct implementation of VAX architecture specification.

7. **Verifiable**: Does the protection system enforce the system's security goals?

The VAX VMM's security goals are embodied in the Bell-LaPadula and Biba information flow models. If Bell-LaPadula was strictly followed, then no software could leak data to an unauthorized user. Similarly, if Biba was strictly followed, then no software would depend on untrusted code or data. In practice, Bell-LaPadula and Biba are both too restrictive, as the designers acknowledge by the addition of system privileges. However, the ability to verify that the system ensures a concrete security goal in the presence of a set of assigned privileges is a challenging task. The VAX VMM system provided no specialized support for understanding the impact of privilege assignments on the resultant information flows. Presumably, these would be manually verified by the administrators of individual deployments.

The design of the VAX VMM system also provided several countermeasures for controlling covert channels. In addition to an informal analysis, the designers also used a technique called the Shared-Resource Matrix [162, 163] to identify storage channels. Many storage channels were eliminated by preallocation of resources, excepting storage channels due to disk arm movement [275] which required a special technique [160]. Covert timing channels are even more difficult to address, and in the course of the VAX VMM project new means were developed to identify covert timing channels [341] and counter these channels by obfuscating timing [140].

11.2.3 VAX VMM RESULT

Despite the diligent efforts of the VAX VMM team, successful pilot deployments, and the A1-assurance preparations, Digital Equipment Corporation (DEC) canceled the project in March 1990. The exact reasons for the cancellation have not been revealed and remain a mystery. However, the reasons why a functional, high security system would be discarded just as it became ready for commercial deployment would be illuminating to those aiming to build high security software in the future.

---

[1] Specifically, the part of the system design called the *formal top-level specification* was modeled and analyzed.

The VAX VMM kernel was stable in early 1988, self-hosting [2] by mid-1988, and supported DEC's VMS and Ultrix-32 operating systems by 1989. An external field test was performed in late 1989, and execution performance was found to be "acceptable."

Undoubtedly, the field test and other business case information was used to determine not to go forward with the VAX VMM as a commercial product. Despite the secrecy, we wonder what aspects of the system caused it not be sold, given acceptable performance. In the rest of this section, we examine some of the challenges faced in building the VAX VMM kernel system that we should consider when constructing future systems.

First, all the device drivers were run in the VMM, which ensured trusted access to hardware, particularly in these days well before an I/O MMUs (see the discussion on I/O MMUs and direct memory addressing in Chapter 6). However, as operating systems must support a large number of devices, new devices are introduced frequently (and often contain bugs [82], and the A1-assurance required detailed code reviews for any new code, this would present a challenge for system maintenance.

Second, the combination of Pascal, PL/1, and a significant amount of assembler (nearly 1/4 of the VMM code) was somewhat unusual for operating systems, circa 1990. Particularly the presence of a significant amount of assembly code, over 11,000 LOC, would present challenges in maintaining A1 assurance.

Third, the VAX multiprocessor hardware introduced high performance covert channels (see Section 5.4) on the shared multprocessor bus. To remedy this problem, the most effective solution given hardware constraints was to use a *fuzzy time* [140] approach whereby certain operations are delayed to prevent an attacker from accurately communicating using such a channel. Delays naturally slow the system's performance, particularly for I/O operations which are the focus of the delays. Although the research paper reports a slowdown of only 5-6%, the performance analysis is not detailed, so some significant performance degradations may have been a concern for commercialization.

11.3 SECURITY IN OTHER VIRTUAL MACHINE SYSTEMS

We briefly examine the state of security in current VM systems. While the initial motivation for VMs was to better leverage physical resources, security has become a recent focus. Current commercial operating systems have failed to provide adequate, manageable security controls, but building a brand new operating system is no longer a practical option. As in the case of the VAX VMM, a new VM system enables the development of new security controls while still being able to execute existing software, and the ability to execute legacy code is a requirement of any system. However, we must construct VM systems in a manner that avoids the problems of operating systems, such as rootkits [170].

---

[2]By *self-hosting*, we mean that the VAX VMM could be used to build new versions of the VAX VMM.

PR/SM   IBM's Processor Resource/Systems Manager [142] (PR/SM, pronounced "prism") is a type 1 hypervisor that is capable of running a variety of operating systems, including various IBM OSes and Linux, on IBM mainframe hardware, now called the *zSeries*. PR/SM enables a security administrator to configure virtual machines such that complete isolation is ensured. That is, PR/SM prevents the sharing of the physical systems I/O resources, so no virtual machine can learn about any virtual machines use of system I/O. PR/SM enables such isolation using an *Interpretive Execution Facility* that only allows virtual machines to execute processor instructions in a controlled manner, which restricts covert channel communications. PR/SM has been evaluated to the Common Criteria EAL 5 in 2005 [14]. Since the focus of the PR/SM product is isolation its evaluation does not aim for MLS protection profiles, such as LSPP (see Chapter 12).

VMware   Although VM systems have been around for many years (e.g., IBM z/VM [143]) and have had a reputation for security for quite some time, the introduction of VMware, a VM system for the ubiquitous x86 processor, brought VM systems to the masses [299]. VMware security is enabled by its dynamic translation of resource requests (CPU, memory, I/O) into virtualized commands that VMware VM monitors can mediate. VMware can both restrict the types of commands that can be performed via this translation and determine the interface for mediation.

When a VMware guest VM executes a storage or network request, the VMware VMM mediates the request and determines how this request will actually be implemented. For network requests, the VMware ESX server defines *virtual ports* which determine the network configuration for that VM (e.g., its MAC address and forwarding tables) independently from its physical platform [46]. A template specification is attached to a VM from which its connectivity can be defined regardless of the host upon which it runs. For storage, traditional I/O requests to a local disk can be converted into a variety of storage system requests [46]. Thus, simpler I/O requests can be authorized and converted into SAN requests, if desired.

The VMware VirtualCenter defines the policy over VMware VM interactions. Resources are grouped into resource pools which partition the CPU and memory resources. Resources pools may be arranged hierarchically and delegated to other subjects. The VMware VirtualCenter uses Windows security controls (see Chapter 7) and roles to define access. The partitioning of resources using pools enables isolation, but if one VM can delegate control of a pool resource to another VM, then a compromise may lead to an information flow that violations the system's security goals. VMware's system design includes a variety of hardening features to protect the VMM's trusted computing base [47], but this does not ensure that a guest VM with control of resource pools cannot be compromised. VMware supports the introspection of guest VMs [106], which may enable the detection of a guest VM compromise from the protected VMM [321]. By running the intrusion detection software (e.g., virus scanners) outside the guest VM protects it from being compromised as well, it does not prevent intrusion or malicious behavior unless the compromise is detectable.

NetTop The first major effort that aimed to leverage VMware for security was the NetTop project [205, 136]. NetTop provides end-to-end secrecy protection using virtual machines for isolation, rather than physical machines. The NetTop system is built on a Type 2 VMware system that uses SELinux [229] (see Chapter 9) as its host operating system. Prior to NetTop, organizations with stringent secrecy requirements used isolated networks to prevent leakage of information. Because of the lack of security in commercial operating systems, these organizations required that distinct machines be connected to each network to prevent leakage. Using NetTop, individual VMs can be isolated on the same physical machine, so a single physical machine can be used to connect to multiple isolated networks.

VMware provides the isolation primitives for NetTop. In addition to storage and memory isolation, VMware supports virtualized VLANs, which limits the set of destination systems to which a VM may send network traffic in a LAN. For some time, network switches have supported the configuration of network partitions in a LAN, such that machines can only send to other members of their VLAN partition. VMware supports the assignment of a virtual machines to distinct VLANs, such that a VM can only send packets to other members of its VLAN partition (other VMs or physical machines). There are different ways to configure VMware VMs to use VLANs [46], but we describe one approach here. Using virtual ports, each VLAN is assigned to virtual port, and each guest VM is assigned to the virtual port as well. The virtual port tags the network frames, so that the VLAN switches can ensure that there is no leakage between VLANs.

Why VMware provides isolation primitives, SELinux is used to ensure that any inter-VM communication is authorized according to a mandatory access control policy. VMware by default uses a discretionary access control policy (based on Windows), so resources can be delegated by users and the VMs. Thus, a compromised VM could delegate rights that would cause an unauthorized leak of information. NetTop uses SELinux in two ways: (1) SELinux access control within a privileged VM defines a *least privilege* [265] policy whereby the permissions of services are contained, even if they are compromised and (2) SELinux defines the resource allotted to VMs by the system, and no delegation is authorized. For example, each virtual machine in NetTop is only authorized to access files with a corresponding label. As a result, a VM in one partition cannot access files in another partition. With the VMware isolation and SELinux policy, NetTop can isolate VMs on the same machine, under the assumption that the trusted computing base cannot be compromised. Even with the least privilege policy, it may be possible for particular trusted services to compromise NetTop should they be compromised.

Xen Xen is another x86 virtual machine system. Xen provides similar isolation guarantees as VMware (e.g., support for VLANs), but the management of inter-VM communication is different. Unlike the discretionary controls of VMware or the SELinux controls within a privileged VM, Xen provides mandatory access control (MAC) at the hypervisor (i.e., VMM) level. The exact implementation of a reference monitor in the Xen hypervisor is an ongoing project, but the aim is similar to the Linux Security Modules (LSM) interface in Linux [342]. Xen isolation, without leveraging its MAC enforcement, has been applied to isolate web applications [66].

The Xen design aimed for virtualization performance originally, but security has also become a focus for Xen. Xen is a Type 1 VMM consisting of two major components: (1) a hypervisor that runs directly on the hardware and (2) a privileged VM that provides I/O and VM configuration support. The Xen hypervisor provides VM communication primitives, mainly aiming at increasing the performance of I/O emulation between the untrusted VMs and the privileged VM which does the actual I/O processing. Unlike the VAX VMM, Xen uses a VM (i.e., a traditional operating system running in a VM, Linux) to provide I/O emulation to the untrusted VMs.

Security for Xen consists of mediating communication between VMs and controlling the distribution of resources (i.e., access to physical devices and memory). To achieve this, two projects have added a reference monitor interface to the Xen hypervisor. Xen sHype is a reference monitor interface that mainly focuses on inter-VM communication control [263]. The Xen communication mechanisms (i.e., *grant tables* and *event channels* [19]) are mediated to ensure that only authorized VMs can communicate. However, the distribution and use of resources is controlled by the privileged VM in an ad hoc manner. Since Xen's privileged VM partitions disk and memory resources and multiplexes network resources among its VMs, no overt communication using these resources is permitted by default. However, this puts a lot of trust in the ordinary OS (Linux) running in the privileged VM to partition resources correctly. Should these resources be shared among VMs in the future, then there sharing must be mediated by trusted software. The second project, the Xen Security Modules (XSM) reference monitor interface mediates both communication and resource distribution [57]. However, now that the reference monitor hooks are in place, it is necessary to determine the policies necessary to utilize these XSM hooks to enforce practical security goals.

11.4   SUMMARY

Virtual machine systems, and more generally separation kernels, provide a layer of abstraction between the operating system and the physical platform. This enables better utilization of hardware, the ability to run multiple operating systems and their applications on one device, and a point of indirection that may be beneficial for security control. Because the complexity of operating systems has prevented them from being effective and manageable arbiters of security, separation kernels and virtual machine monitors are seen as a layer where security guarantees can be practically enforced. While these systems have been around for a while and have historically supported security, secure VMMs (i.e., in the manner required in Chapter 2) are not readily available. The VAX VMM defined a formally rigorous design and implementation of a secure VMM system, but it was never released as a product. The VAX VMM design demonstrated that VMM security depends on control of all sensitive commands, including those performed by I/O resources. But, this created a conflict, as the number and variety of drivers tends to overwhelm the code management and testing required for formal assurance. MILS separation kernel systems are also being developed, but are specialized systems not yet leveraged by mainstream computing. VMware and Xen are available VMMs that are being extended to support the enforcement of security, but they emulate I/O in an ordinary operating system. The trade-off between what function belongs in the VMM, and what function can be performed in ordinary operating systems with sufficient security guarantees is an ongoing source of debate.

CHAPTER 12

System Assurance

The aim of *system assurance* is to verify that a system enforces a desired set of security goals. For example, we would like to know that a new operating system that we are developing can protect the secrecy of one group of users' data from another group. First, we would like to know that the operating system mechanisms (e.g., reference monitor) and policies (e.g., multilevel security) are appropriate to enforce the goal. Second, we would like to know whether the operating system implementation correctly implements intended mechanisms and policies. System assurance describes both *what* determines reasonable goal and what is a satisficing implementation, and system assurance also describes *how* a secure operating system should be built and maintained.

The development of Multics, see Chapter 3, and the subsequent security kernel systems, see Chapter 6, demonstrated to the security community that building a secure operating system was a difficult undertaking. The security kernel approach emerged as the appropriate technique for constructing secure system, see Section 6.1. Three specific tasks were identified [10, 108]. First, the system development team must clearly define the security goals for their system. Second, the system development team must construct a system design in a such a way that its security properties can be verified, formally if possible. Third, the system development team must implement the kernel in a manner that can be traced to the verified design, automatically if possible. This approach to system development motivation the construction of several design verification tools(e.g., [48, 118, 79, 78, 331]).

Using the security kernel approach as a guide, the US Department of Defense (DoD) developed a set of standards for the security requirements of systems and evaluation procedures for verifying these requirements, called the *Rainbow Series* [222]. These standards cover security function ranging from passwords and authentication to recovery and audit to access control and system configuration. Further, there are standards for procedures including documentation, procurement, facilities management, etc. The standards covering operating systems, called the Trusted Computer System Evaluation Criteria (TCSEC) or the "Orange Book" [304], have generated the most discussion and probably the most use. The Orange Book defines a progressively-more-secure sequence of requirements for operating systems. The Orange Book combines desired security features with the tasks to verify correct implementation of those features into a set of assurance classes. Ultimately, this approach was found to be unnecessarily constraining, as the levels of security function and assurance may not coincide. In the succeeding assurance approach, called the Common Criteria [60], evaluation criteria are organized into distinct components: the security function into the *target of evaluation* (TOE) and evaluation effort into the *evaluation assurance levels* (EALs). Thus, a system's particular security features can be assured to different levels depending on its design, development, documentation, configuration, testing, etc.

In this chapter, we describe the Orange Book and Common Criteria assurance methodologies and their impact on operating system design and implementation.

12.1   ORANGE BOOK

The TCSEC or Orange Book was developed by the US DoD's Computer Security Center which was formed in 1981 [304]. The main focus of the center was to encourage the development of secure operating systems by vendors. However, as discussed throughout the book, building a secure operating system is a challenging task, and by this time the commercial market had largely moved away from the features of secure operating systems (e.g., Multics [237] and security kernels, such as Scomp [99] and GEMSOS [290]) to the function and flexibility of UNIX systems which did not focus on security (see Chapter 4). Thus, the center needed to not only define how to build such systems, but also encourage the development of secure operating systems.

The Orange Book defines two sets of requirements: (1) specific security feature requirements and (2) assurance requirements. The *specific security feature requirements* define the operating system features that are necessary to enforce the security requirements. The *assurance requirements* specify the effort necessary to verify the correct implementation of the specific security features.

The Orange Book is primarily known for its definition of *assurance classes* that dictate specific security feature and assurance requirement combinations. Ultimately, these classes were organized into *divisions* with similar assurance requirements (A through D). Each class defines four categories of requirements for their systems: (1) the *security policy* model, including the administration of policies described in the model and the labeling of system resources; (2) the level of *accountability* for system actions, including authentication of individual subjects and audit of system actions; (3) the degree of *operational assurance* that the system behaves as expected, including the implementation and maintenance of the system; and (4) the *documentation* provided to support the design, implementation, assurance, and maintenance of the system. The first two categories describe the specific security feature requirements that dictate the security function intended by the system design. The second two categories specify assurance requirements that determine whether the implementation satisfies the intended design. Each class specifies its minimal requirements for these categories.

We describe Orange Book classes below, starting with the lowest security classes (i.e., least assurance of security enforcement).

D – Minimal Protection   Class D is reserved for systems that have been evaluated, but fail to meet the requirements of any higher security class.

The next set of classes support discretionary access control security policies. These classes are grouped into a division called *Discretionary Protection* (division C) where only discretionary mechanisms are necessary to evaluated be at one of these classes. Other features and assurance requirements distinguish the classes.

C1 – Discretionary Security Protection   A Class C1 system requires a discretionary access control (DAC) model specifying the permissions of named users to named objects. Users must authenticate themselves prior to performing any system actions. System actions are performed by hardware-protected domains whose rights are associated with the authenticated users. Assurance requires testing that there are no obvious ways to bypass these controls. Basic documentation is required.

This class defines a basic DAC system with hardware protection of processes and user authentication.

C2 – Controlled Access Protection   A Class C2 system provides the DAC model of Class C1 where the rights may be specified at the granularity of a single user, and where administration is authorized. Authentication shall be based on a secret (e.g., a password) that is protected from access by other users. Audit and object reuse are introduced by this class. Audit of a specific set of events is defined, and such auditing requires a protected log. Object reuse means that when an object is provided, such as a file, no data from its previous use is accessible. Assurance requires testing for obvious flaws and obvious bypass. Documentation for users, facilities, design, and testing is required.

C2 is the evaluation level for most discretionary systems, such as Windows and UNIX systems, historically. Access control is based on a user identity, and passwords are protected from access. As of 1996, five operating systems were assured at this class, including Windows NT 3.5 Service Pack 3, albeit without networking.

The next division, called *Mandatory Protection*, defines classes of systems that provide mandatory access control (MAC).

B1 – Labeled Security Protection   A Class B1 system provides the DAC, audit, and object reuse features of Class C2 plus a MAC model where each named subject and object is associated with a sensitivity label, corresponding to a multilevel security (MLS) policy (see Section 5.2). These labels must be integrity-protected and exported with the data when it leaves a system (e.g., by device or via a printed page). Authentication must identify a user by sensitivity level for authorization. Assurance now requires that the security mechanisms of the system must work as claimed in the system documentation. An examination of the design documentation, source code, and object code are necessary to prove this. The documentation must support such testing through detailed descriptions of the security policy model and protection mechanisms, including a justification how they satisfy the model.

B1 is the class in which mandatory access control is introduced. Also, at this stage detailed testing that the documentation and source code implement the intended security features is required. Seven operating systems were evaluated at this class, including Trusted Solaris V1.1 Compartmented Mode Workstation (see Chapter 8).

B2 – Structured Protection   A Class B2 system extends the B1 class by requiring enforcement on access to all subjects and objects (i.e., not just named ones) and covert channel protections (see Section 5.4). At this stage, authentication requires a trusted path (see Section 7.5). Also, requirements on the trusted computing base (TCB) design are added. The protection-critical part of the TCB must be identified, and its interface must be well-defined. At this point, the TCB must be shown to be "relatively resistant to penetration."

B2 is the class at which covert channels are first mentioned. This introduces a new significant and complex design and evaluation task to the assurance process. The evaluation of the TCB is also much more detailed. The design specification includes a "descriptive top-level specification" that includes an accurate description of the TCB interface and its exceptions, error messages, and effects. Only two systems were evaluated at this class, Trusted Xenix 3.0 and 4.0 [111].

B3 – Security Domains   A Class B3 system extends the B2 class by requiring that the TCB satisfy the reference monitor concept in Definition 2.6. The TCB design and implementation are directed toward minimal size and minimal complexity. The system is expected to be "highly resistant to penetration." At this point, the audit subsystem must be able to record all security-sensitive events.

At this point, we see the requirements for a secure operating system, but without the formal verification necessary for Division A assurance below. As of 1996, only one system was assured at B3, the XTS-300 system [21] that was a successor to the Scomp system discussed in Section 6.2.

A1 – Verified Design   A Class A1 system is functionally equivalent to a B3 system, but the evaluation of this system must be derived from a formal design specification. Unlike the other classes, the assurance of a Class A1 system is developmental in that the design and implementation of the system follow from a formal top-level specification (FTLS).

A Class A1 system must meet the following five requirements:

1. A formal model of the security policy must be documented and include a mathematical proof that the model is consistent with the policy.

2. An FTLS must specify the functions that the TCB performs and the hardware/firmware support for separate execution domains.

3. The FTLS of the TCB must be shown to be consistent with the formal model of the security policy.

4. The TCB implementation must be consistent with the FTLS.

5. Formal analysis techniques must be used to identify and analyze covert channels. The continued existence of covert channels in the system must be justified.

A Class A1 system is a secure operating system that has been semi-formally verified to satisfy the reference monitor guarantees. Building a Class A1 system requires diligence from the outset of the development process to build and verify the formal specifications of the system design (FTLS), the security policy, and the TCB of the system. No Class A1 operating systems are commercially available, although custom systems, such as BLACKER [330] based on GEMSOS [290], have been evaluated at A1, and the VAX/VMM system (see Section 11.2) was designed and implemented to be evaluated at Class A1.

Beyond Class A1   The Class A1 requirements include exceptions that permit informal verification where no formal analysis tool exist. In particular, research in the 1980s indicated that formal verification of security properties down to the level of the source code may be possible, so the Beyond Class A1 left open the possibility that security requirements could be verified more precisely. Formal verification tools that can prove the satisfaction of the variety of security properties required for large source code bases, such as operating systems, have not become practical. As a result, no further classes were defined.

12.2 COMMON CRITERIA

While the Orange Book defined targets for security features and assurance, the linking of features and assurance requirements became cumbersome. A detailed formal assurance may be performed for systems that do not provide all of the features in B3 (e.g., covert channel protection). Also, the six classes of security features themselves were constraining. The market may determine combinations of security features that do not neatly fit into one of these classes. Out of this desire for a more flexible assurance approach, the Common Criteria emerged [60, 61].

In the early 1990s, independent approaches for system assurance were be developed in Europe and Canada. The Information Technology Security Evaluation Criteria (ITSEC) version 1.2 was released in 1991 as a joint standard used by France, Germany, the Netherlands, and the United Kingdom [147]. ITSEC defines a set of criteria for evaluating a system, called the *target of evaluation*, to verify the presence of a set of security features and to verify its defenses against a comprehensive set of penetration tests. The set of security features implemented by a system need not conform to a specific TCSEC class, but rather can be defined as part of the evaluation, called the *security target*. The amount of evaluation effort determines a confidence level in the target of evaluation, called *evaluation levels* that range from E0 (lowest) to E6 (highest).

In Canada, the Canadian Trusted Computer Product Evaluation Criteria (CTCPEC) defined its evaluation approach in 1993 [41, 194]. CTCPEC leverages the security feature requirements of the Orange Book while incorporating the notion of distinct assurance levels of the ITSEC approach.

Inspired by facets of each of these evaluation approaches, the Common Criteria approach was developed [60]. Like the ITSEC and CTCPEC approaches, the Common Criteria separates the assurance effort from the security features being assured. Like the Orange Book, the Common Criteria defines security feature configurations that would lead to a meaningful assurance for developers and consumers.

12.2.1 COMMON CRITERIA CONCEPTS

An overview of the Common Criteria approach is shown in Figure 12.1. The Common Criteria

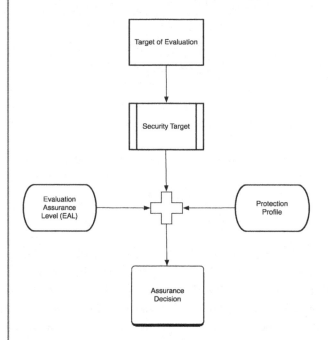

Figure 12.1: Common Criteria Overview: A *target of evaluation* is converted to a *security target* that defines the features and assurance being evaluated. The *protection profile* specifies the security features required and the *assurance levels* define a set of possible assurance requirements for those features.

consists of four major concepts: (1) the *target of evaluation* (TOE) which is the system that is the subject of evaluation; (2) the *protection profile* (PP) which specifies a required set of security features and assurance requirements for systems to satisfy that profile; (3) the *security target* (ST) which defines the functional and assurance measures of the TOE security that are to be evaluated, perhaps satisfying one or more PPs; and (4) the *evaluation assurance level* (EAL) which identifies the level of assurance that the TOE satisfies the ST. In a Common Criteria evaluation, the TOE's security threats, security objectives, features, and assurance measures are collected in an ST for that system. The ST may claim to satisfy the requirements of one or more PPs which defines a set of requirements on the ST. The PPs typically indicate a minimum EAL that can be assured using its assurance requirements, but the ST may exceed these assurance requirements to satisfy the requirements of a higher EAL. We examine these concepts in more detail below.

The protection profiles (PPs) are generally derived from the popular TCSEC classes. For discretionary access control (DAC) systems, the *Controlled Access Protection Profile* (CAPP) was developed that corresponds to TCSEC Class C2 [230]. The functional requirements of a CAPP system correspond to a Class C2 system, and CAPP also defines assurance requirements for a typical system with a minimum EAL 3 assurance required (see below for EAL definitions). For mandatory access control (MAC) systems, the *Labeled Security Protection Profile* (LSPP) was developed [231]. It corresponds to TCSEC Class B1. LSPP also defines assurance requirements that require a minimum of EAL3 assurance, although a higher level of assurance is expected.

Level and TCSEC Map	Requirements
EAL1	functionally tested
EAL2 (C1: low)	structurally tested
EAL3 (C2/B1: moderate)	methodically tested and checked
EAL4 (C2/B1: medium)	methodically designed, tested and reviewed
EAL5 (B2: high)	semiformally designed and tested
EAL6 (B3: high)	semiformally verified design and tested
EAL7 (A1: high)	formally verified design and tested

Figure 12.2: Common Criteria Evaluation Assurance Levels

The evaluation assurance levels (EALs) and their abstract test and verification requirements are listed in Table 12.2. The EAL levels indicate identifiable assurance that the ST is implemented correctly by the TOE system. It is important to note that the assurance level does not indicate that a system has more security features, but rather whether the system's design and implementation assures that the support security features are correct. Both the security features and assurance must be assessed to determine whether the system satisfies the requirements of a secure operating system (Definition 2.5 in Chapter 2).

Table 12.2 also includes an indication of high, medium, moderate, and low assurance for the levels. This categorization is not universally agreed upon, but the important thing to take away is that there are significant assurance differences between the aggregations. For example, EAL3 is the first level to require a methodical testing, including grey box testing and a search for vulnerabilities. Also, a high-level design description is required. However, EAL4 assurance requires a much higher effort on the part of the developer for a low-level design and a detailed vulnerability test. EAL5-7 are all considered as high assurance as the levels all require some application of formal methods to the assurance. Only 2 products on the Common Criteria's List of Evaluated Products have been evaluated at EAL7 (two data diode devices for restricting the flow of secret data from Tenix) and none have been evaluated at EAL6. For operating systems, only the Processor Resource/System Manager (PR/SM) system from IBM and the XTS-400 system from BAE systems are evaluated at EAL5.

Security Targets (STs) are constructed from the TOE system using a set of predefined components. There are functional components that cover security features, such as for audit, cryptographic support, user data protection (access control), and authentication. There are also assurance components that cover the extent to which the ST features can be evaluated, including configuration management, system delivery and operation (deployment), documentation, and testing. The assurance components also include ones for assuring that PPs are satisfied and that the ST itself is complete, consistent, and technically sound.

The components form the foundation of STs and PPs, and as such they are low-level. For example, CAPP uses *functional components* [58] for User Data Protection for discretionary access control policy (indicated by a component identifier, FDP_ACC.1), discretionary access control function (FDP_ACF.1), and object reuse (FDP_RIP.2). User Data Protection forms a class of requirements, identified by the name FDP. Each of these components that were chosen belong to different families (FDP_ACC, FDF_ACF, and FDP_RIP, respectively). LSPP builds on the CAPP requirements for User Data Protection by adding functional components for mandatory access control (FDP_IFC.1) and label import and export (FDP_ITC.2 and FDP_ETC.2), among others. Thus, the PP definitions consist of a set of functional components organized by their classes.

Similarly, STs would be defined from the same pool of functional components, but they must also be defined in terms of assurance components in order to be evaluated against the specified EAL. Assurance components refer to classes of requirements in the areas of configuration management (ACM), delivery and operation (ADO), life cycle support (ALC), vulnerability assessment (AVA), etc. The EAL definition includes a set of components from these classes.

12.2.2 COMMON CRITERIA IN ACTION

Recently, RedHat Linux Enterprise Linux Version 5 was assured at EAL4 for LSPP. What did it take to perform this evaluation? First, the Linux system formed the TOE of evaluation, so an ST for this system was constructed. Building an ST for achieving LSPP at EAL4 requires by collecting all the components required by LSPP [231] and all the components required by EAL4 [59] into a single ST specification. LSPP requires nearly 40 functional requirement components and nearly 20 assurance requirement components. The functional requirement components include the obvious ones for multilevel security (MLS), which is at the heart of the profile, but also components for export/import of labeled data, among others. The assurance requirement components for LSPP include those necessary to describe the system design (a high-level design for LSPP) and manage the deployment and maintenance of the system. As the assurance level does not impact function, EAL4 adds only assurance requirements to those required for LSPP. LSPP is designed to support EAL3 evaluation, so the additional requirements of EAL4 evaluation above EAL3 must be added to the ST. There are 7 additional assurance requirements for EAL4 over EAL3. While this is a small number of components, these require significant additional evaluation effort. These additional components include a complete functional specification, an implementation representation of the system, module-level design, and a focused vulnerability analysis.

Once the Linux ST is defined, the TOE must be evaluated against these requirements. For each of the nearly 70 requirements, a set of *elements* must be fulfilled. For example, a complete functional specification (ADV_FSP.4) has 10 elements. There are elements related to the development of the system, verification of the system's content and presentation, and evaluation of the quality of the other elements. First, two elements must be provided by the developer: (1) a functional specification and (2) a description linking the functional specification to all the security functional requirements. A case must be made as to how the Linux TOE satisfies those requirements. These elements define the system structure and how it implements the required security functions.

Next, there are six content and presentation elements that specify how the system satisfies its security requirements. These include: (1) demonstrating that the functional specification completely represents the trusted computing base of the system, called the *target of security function* (TSF); (2) the purpose and method of each interface in the TSF; (3) a description of all the parameters for each interface in the TSF; (4) all the actions performed based on interface invocation of the TSF; (5) all error messages that may result due to an interface invocation of the TSF; and (6) a description linking the security functional requirements to the interfaces of the TSF. Cumulatively, these describe how the system's trusted computing base is invoked and how it implements the required security function when invoked.

Lastly, the evaluator must confirm that the specifications above are complete and correct. There are two elements required: (1) the evaluator shall confirm that the information provided meets all requirements for content and presentation of evidence and (2) the evaluator shall determine that the functional specification is an accurate and complete instantiation of the security function requirements.

Note that this is just one of the nearly 70 requirements that needs to be fulfilled as part of the evaluation. While this is one of the most demanding elements, the level of effort necessary to fulfill each of the elements is significant. Many pages of documentation are generated in the assurance process, and resulting costs are impressive.

In the end, the degree of effort taken in an evaluation process should have some positive effect on the quality of the target system. However, even the effort put into verifying the functional specification does not prove correctness in any concrete sense. Notice that the requirements above only require a manual examination of code and an informal description of how the security requirements are implemented.

As EAL4 aims for medium assurance, this level of analysis is appropriate, but it is generally accepted that general-purpose, commercial operating systems, such as Linux, BSD, and Windows, will never be evaluated at a higher assurance level due to their complexity and the limits of their development processes. For higher assurance, the development process must be tightly controlled, such that the formal specifications can be generated and tested against the code. In the future, we need assurance techniques and tools that enable effective development while constructing the necessary case for source code level assurance. These challenges have been with the security community since the evaluation process was proposed, but we still have a ways to go.

12.3 SUMMARY

System assurance is the process of verifying that a system enforces a desired set of security goals. The TCSEC or Orange Book was developed in the early 1980s to define how to verify security in operating systems. The Orange Book defined a set of classes that specified distinct security and assurance requirements for operating systems. The Orange Book notion of system assurance gained acceptance, if only because the US DoD mandated the use of assured systems.

However, the European and Canadian and eventually American security communities saw the need to separate the security function of a system from the assurance that it truly implements that function. The Orange Book notions of security function were extended and combined with the assurance approaches developed in Europe (ITSEC) and Canada (CTCPEC) to create the Common Criteria approach.

Many more evaluations have now been performed using the Common Criteria than were ever performed under the Orange Book, resulting a worldwide approach to assuring systems. Performing a system evaluation is still strongly motivated by government markets, and the cost of performing even medium-level assurance for most systems is prohibitive. Further, the current, commercial development process precludes high assurance of systems. The weak link in assurance has been and continues to be techniques and tools to build systems whose security can be verified in a mostly, automated fashion. While this problem is intractable in general, work continues on developing usable, formal verification techniques (e.g., NICTA [81]).

Bibliography

[1] Solaris Trusted Extensions Developer's Guide.
http://docs.sun.com/app/docs/doc/819-7312, 2008.

[2] M. Abadi, E. Wobber, M. Burrows, and B. Lampson. Authentication in the Taos Operating System. In *Proceedings of the 14th ACM Symposium on Operating System Principles*, pp. 256–269, Asheville, NC, 1993. DOI: 10.1145/168619.168640

[3] A. Acharya and M. Raje. MAPbox: Using parameterized behavior classes to confine untrusted applications. In *Proceedings of the 9th USENIX Security Symposium*, August 2000.

[4] W. B. Ackerman and W. W. Plummer. An implementation of a multiprocessing computer system. In *Proceedings of the First ACM Symposium on Operating Systems Principles*, 1967. DOI: 10.1145/800001.811666

[5] Acsec corporation, 2008. http://www.aesec.com/.

[6] A. Alexandrov, P. Kmiec, and K. Schauser. Consh: A confined execution environment for internet computations. In *Proceedings of the 1999 USENIX Annual Technical Conference*, 1999.

[7] J. Alves-Foss, W. S. Harrison, P. Oman, and C. Taylor. The MILS architecture for high assurance embedded systems. *International Journal of Embedded Systems*, 2007. In press. DOI: 10.1504/IJES.2006.014859

[8] AMD I/O Virtualization Technology (IOMMU) Specification. Technical Report 34434, Advanced Micro Devices, Inc.
http://www.amd.com/us-en/assets/content_type/white_papers_and_tech_docs 34434.pdf, February 2007.

[9] B. Ames. Real-time software goes modular. *Military & Aerospace Electronics*, 14(9), 2003.

[10] S. A. Ames, M. Gasser, and R. R. Schell. Security Kernel Design and Implementation: An Introduction. *IEEE Computer*, 16(7):14–22, 1983. DOI: 10.1109/MC.1983.1654439

[11] J. P. Anderson. Computer security technology planning study. Technical Report ESD-TR-73-51, The MITRE Corporation, Air Force Electronic Systems Division, Hanscom AFB, Badford, MA, 1972.

[12] M. Anderson, R. D. Pose, and C. S. Wallace. A password capability system. *The Computer Journal*, 29(1):1–8, February 1986. DOI: 10.1093/comjnl/29.1.1

[13] Security starts with your operating system. `http://www.argus-systems.com/home3.shtml`, 2008.

[14] Evaluation of IBM PR/SM z/Series990/890. atsec News Release at `http://www.atsec.com/01/news-article-63.html`, 2005.

[15] L. Badger, D. F. Sterne, D. L. Sherman, K. M. Walker, and S. A. Haghighat. A domain and type enforcement UNIX prototype. In *Proceedings of the 5th USENIX Security Symposium*, 1995. DOI: 10.1109/SECPRI.1995.398923

[16] L. Badger, D. F. Sterne, D. L. Sherman, K. M. Walker, and S. A. Haghighat. Practical Domain and Type Enforcement for UNIX. In *SP '95: Proceedings of the 1995 IEEE Symposium on Security and Privacy*, p. 66, IEEE Computer Society, Washington, DC, 1995. DOI: 10.1109/SECPRI.1995.398923

[17] A. Baliga, P. Kamat, and L. Iftode. Lurking in the Shadows: Identifying systemic threats to kernel data. In *Proceedings of the 2007 IEEE Symposium on Security and Privacy*, pp. 246–251, May 2007. DOI: 10.1109/SP.2007.25

[18] T. Ball and S. Rajamani. The SLAM toolkit: Debugging system software via static analysis. In *Proceedings of the ACM Conference on Principles of Programming Languages*, January 2002.

[19] P. Barham, B. Dragovic, K. Fraser, S. Hand, T. Harris, A. Ho, R. Neugebauer, I. Pratt, and A. Warfield. Xen and the art of virtualization. In *Proceedings of the Symposium on Operating Systems Principles (SOSP)*, October 2003. DOI: 10.1145/945445.945462

[20] Bastille Linux. `http://www.bastille-linux.org/`.

[21] DigitalNet - Solutions - Information Assurance. `http://www.digitalnet.com/solutions/information_assurance/xts300sol_ste.htm\#hadg`, 2008.

[22] XTS-400 Trusted Computer System, from BEA Systems. `http://www.baesystems.com/ProductsServices/bae_prod_csit_xts400.html`, 2008.

[23] D. E. Bell and L. J. LaPadula. Secure computer system: Unified exposition and Multics interpretation. Technical Report ESD-TR-75-306, Deputy for Command and Management Systems, HQ Electronic Systems Division (AFSC), L. G. Hanscom Field, Bedford, MA, March 1976. Also, MITRE Technical Report MTR-2997.

[24] M. Bellis. Inventors of the modern computer. `http://inventors.about.com/library/weekly/aa033099.htm`.

[25] J. L. Berger, J. Picciotto, J. P. L. WOodward, and P. T. Cummings. Compartmented Mode Workstation: Prototype highlights. *IEEE Transactions on Software Engineering*, 16(6):608–618, June 1990. DOI: 10.1109/32.55089

[26] V. Berstis. Security and protection of data in the IBM System 38. In *Proceedings of 7th Symposium on Computer Architecture*, pp. 245–252, May 1980. DOI: 10.1145/800053.801932

[27] K. J. Biba. Integrity considerations for secure computer systems. Technical Report MTR-3153, MITRE, April 1977.

[28] A. D. Birrell, A. Hisgen, C. Jerian, T. Mann, and G. Swart. The Echo distributed file system. Technical Report 111, Digital Systems Research Center, September 1993.

[29] M. Bishop. *Computer Security: Art and Science*. Addison-Wesley, 2002.

[30] M. Bishop and M. Digler. Checking for race conditions in file accesses. *Computer Systems*, 9(2), Spring, 1996.

[31] M. Bishop and L. Snyder. The transfer of information and authority in a protection system. In *Proceedings of the 7th ACM Symposium on Operating System Principles*, pp. 45–54, 1979. DOI: 10.1145/800215.806569

[32] W. E. Boebert. On the inability of an unmodified capability machine to enforce the *-property. In *Proceedings of the 7th DoD/NBS Computer Security Conference*, pp. 291–293, September 1984.

[33] W. E. Boebert and R. Y. Kain. A practical alternative to hierarchical integrity policies. In *Proceedings of the 8th National Computer Security Conference*, 1985.

[34] W. E. Boebert, R. Y. Kaln, W. D. Young, and S. A. Hansohn. Secure Ada Target: Issues, system design, and verification. In *Proceedings of the 1985 IEEE Symposium on Security and Privacy*, May 1985. DOI: 10.1109/SP.1985.10022

[35] M. Branstad, H. Tajalli, and F. L. Mayer. Security issues of the Trusted Mach system. In *Proceedings of the 1988 Aerospace Computer Security Applications Conference*, pp. 362–367, December 1988. DOI: 10.1109/ACSAC.1988.113334

[36] M. Branstad, H. Tajalli, F. L. Mayer, and D. Dalva. Access mediation in a message passing kernel. In *Proceedings of the 1989 IEEE Symposium on Securityband Privacy*, 1989. DOI: 10.1109/SECPRI.1989.36278

[37] D. F. C. Brewer and M. J. Nash. The Chinese Wall security policy. In *Proceedings of the IEEE Symposium on Security and Privacy*, 1989. DOI: 10.1109/SECPRI.1989.36295

[38] E. Brewer, P. Gauthier, I. Goldberg, and D. Wagner. Basic flaws in internet security and commerce. http://www.cs.berkeley.edu/~daw/papers/endpoint- security.html.

[39] E. Bugnion, S. Devine, K. Govil, and M. Rosenblum. Disco: Running commodity operationg systems on scalable multiprocessors. *ACM Transactions on Computer Systems*, 15(4):412–447, November 1997. DOI: 10.1145/265924.265930

[40] C-R.Tsai, V. D. Gligor, and C. S. Chandersekaran. A formal method for the identification of covert storage channels in source code. In *Proceedings of the 1987 IEEE Symposium on Security and Privacy*, 1987. DOI: 10.1109/SP.1987.10014

[41] The Canadian Trusted Computer Product Evaluation Criteria. Canadian System Security Centre, Communications Security Establishment, Government of Canada, January 1993.

[42] M. Carson, et al. Secure window systems for UNIX. In *Proceedings of the USENIX Winter 1989 Conference*, January 1989.

[43] J. Carter. Using GConf as an example of how to create a userspace object manager. In *Proceedings of the 2007 SELinux Symposium*. Available at http://selinux-symposium.org/2007/agenda.php, March 2007.

[44] S. Chari and P. Cheng. Bluebox: A policy-driven, host-based intrusion detection system. *ACM Transaction on Infomation and System Security*, 6:173–200, May 2003. DOI: 10.1145/762476.762477

[45] J. S. Chase, H. M. Levy, M. J. Feeley, and E. D. Lazowska. Sharing and protection in a single-address-space operating system. *ACM Transactions on Computer Systems*, 12(4):271–307, 1994. DOI: 10.1145/195792.195795

[46] C. Chaubal. Security design of the VMware Infrastructure 3 architecture. VMware Inc., document at http://www.vmware.com/pdf/vi3_security_architecture_wp.pdf, 2007.

[47] C. Chaubal. VMware Infrastructure 3 security hardening. VMware Inc., document at http://www.vmware.com/pdf/vi3_security_hardening_wp.pdf, 2007.

[48] M. H. Cheheyl and et al. Verifying security. *ACM Computing Surveys*, 13(3):279–339, 1981.

[49] H. Chen, D. Dean, and D. Wagner. Model checking one million lines of C code. In *Proceedings of the 11th ISOC Network and Distributed Systems Security Symposium (NDSS'04)*, pp. 171–185, February 2004.

[50] A. Chitturi. Implementing mandatory network security in a policy-flexible system, April/June 1998. University of Utah, Master's Thesis.

[51] J. Chow, B. Pfaff, T. Garfinkel, K. Christopher, and M. Rosenblum. Understanding data lifetime via whole system simulation. In *Proceedings of the 13th USENIX Security Symposium*, 2004.

[52] J. Chow, B. Pfaff, T. Garfinkel, and M. Rosenblum. Shredding your garbage: Reducing data lifetime through secure deallocation. In *Proceedings of the 14th USENIX Security Symposium*, 2005.

[53] Commercial IP Security Option (CIPSO 2.2). http://sourceforge.net/docman/display_doc.php?docid=34650&group_id=174379, July 1992.

[54] D. D. Clark and D. Wilson. A comparison of military and commercial security policies. In *1987 IEEE Symposium on Security and Privacy*, May 1987. DOI: 10.1109/SP.1987.10001

[55] E. Cohen. Information transmission in computational systems. *ACM SIGOPS Operating Systems Review*, 11(5):133–139, 1977. DOI: 10.1145/1067625.806556

[56] E. Cohen and D. Jefferson. Protection in the Hydra operating system. In *Proceedings of the Fifth ACM Symposium on Operating Systems Principles*, pp. 141–160, 1975. DOI: 10.1145/800213.806532

[57] G. Coker. Xen Security Modules (XSM). http://www.xen.org/files/xensummit_4/xsm-summit-041707_Coker.pdf. These are presentation slides from the 2007 Xen Summit, April 2007.

[58] Part 2: Security functional components. http://www.commoncriteriaportal.org/files/ccfiles/CCPART2V3.1R2.pdf. Version 3.1, Revision 2, September 2007.

[59] Part 3: Security assurance components. http://www.commoncriteriaportal.org/files/ccfiles/CCPART3V3.1R2.pdf. Version 3.1, Revision 2, September 2007.

[60] Common Criteria—the Common Criteria portal. http://www.commoncriteriaportal.org, 2008.

[61] Official CC/CEM versions. http://www.commoncriteriaportal.org/thecc.html. Latest version at time of writing is Version 3.1, Revision 2, 2008.

[62] F. J. Corbató and V. A. Vyssotsky. Introduction and overview of the Multics System. In *Proceedings of the 1965 AFIPS Fall Joint Computer Conference*, 1965.

[63] D. C. Cosserat. A capability oriented multi-processor system for real-time applications. In *Proceedings of the 1972 International Conference on Computer Communications*, October 1972.

[64] C. Cowan, C. Pu, D. Maier, H. Hinton, J. Walpole, P. Bakke, S. Beattie, A. Grier, P. Wagle, and Q. Zhang. Stackguard: Automatic adaptive detection and prevention of buffer-overflow attacks. In *Proceedings of the 7th USENIX Security Symposium*, 1998.

[65] R. Cox, E. Grosse, R. Pike, D. Presotto, and S. Quinlan. Security in Plan 9. In *Proceedings of the 11th USENIX Security Symposium*, August 2002.

[66] R. S. Cox, J. G. Hansen, S. D. Gribble, and H. M. Levy. A safety-oriented platform for web applications. In *Proceedings of the 2006 IEEE Symposium on Security and Privacy*, May 2006. DOI: 10.1109/SP.2006.4

[67] L. A. Cox, Jr. and R. R. Schell. The structure of a security kernel for a Z8000 multiprocessor. In *Proceedings of the 1981 IEEE Symposium on Security and Privacy*, pp. 124–129, April 1981. DOI: 10.1109/SP.1981.10015

[68] CP/M main page. http://www.seasip.demon.co.uk/Cpm/index.html.

[69] R. J. Creasy. The origin of the VM/370 time-sharing system. *IBM Journal of Research and Development*, 25(5), 1981.

[70] D. Denning. A lattice model of secure information flow. *Communications of the ACM*, 19(5):236–242, 1976. DOI: 10.1145/360051.360056

[71] D. Denning. *Cryptography and Data Security*. Addison-Wesley, 1982.

[72] J. B. Dennis and E. C. Van Horn. Programming semantics for multiprogrammed computations. *Communications of the ACM*, 9(3):143–155, 1966. DOI: 10.1145/365230.365252

[73] K. D. Dent. *Postfix: The Definitive Guide*. O'Reilly Media, 2003.

[74] J. Dias. A guide to Microsoft Active Directory (AD) design. Technical Report UCRL-MA-148650, Lawrence Livermore National Laboratory, May 2002.

[75] The first PC operating system. http://www.digitalresearch.biz/CPM.HTM.

[76] E. W. Dijkstra. The structure of the "THE"-multiprogramming system. *Communications of the ACM*, 11(5):341–346, 1968. DOI: 10.1145/363095.363143

[77] J. Dike. The User-mode Linux Kernel Home Page. http://user-mode-linux.sourceforge.net/, 2008.

[78] S. Eckmann. Ina Flo: The FDM Flow Tool. In *Proceedings of the 10th National Computer Security Conference*, pp. 175–182, Sept 1987.

[79] S. Eckmann and R. A. Kemmerer. Inatest: An interactive environment for testing formal specifications. *Software Engineering Notes*, 10(4):17–18, 1985. DOI: 10.1145/1012497.1012504

[80] N. Edwards, J. Berger, and T. H. Choo. A secure Linux platform. In *ALS '01: Proceedings of the 5th Annual Conference on Linux Showcase & Conference*, 2001.

[81] K. Elphinstone, G. Klein, P. Derrin, T. Roscoe, and G. Heiser. Kernel development for high assurance. In *Proceedings of the 11th Workshop on Hot Topics in Operating Systems*, 2007.

[82] D. Engler, B. Chelf, A. Chou, and S. Hallem. Checking system rules using system-specific programmer-written compiler extensions. In *Proceedings of the 4th Symposium on Operating System Design and Implementation*, pp. 1–16, December 2000.

[83] D. R. Engler, M. F. Kaashoek, and J. O'Toole. Exokernel: An operating system architecture for application-level resource management. In *Proceedings of the Fifteenth Symposium on Operating Systems Principles*, pp. 251–266, 1995. DOI: 10.1145/224056.224076

[84] System call interception whitepaper. Entercept Security Technologies at http://www.entercept.com/whitepaper/systemcalls/.

[85] J. Epstein and J. Picciotto. Trusting X: Issues in building trusted X window systems -or- what's not trusted about X? In *Proceedings of the 14th Annual National Computer Security Conference*, October 1991. DOI: 10.1109/CSAC.1991.213019

[86] J. Epstein, et al. Evolution of a Trusted B3 Window System prototype. In *Proceedings of the 1992 IEEE Symposium on Research in Security and Privacy*, May 1992. DOI: 10.1109/RISP.1992.213258

[87] F. J. Corbató et al. *The Compatible Time Sharing System: A Programmer's Guide*. MIT Press, first edition, 1963.

[88] F-secure virus descriptions: Code red. http://www.f-secure.com/v-descs/bady.shtml.

[89] R. S. Fabry. Capability-based addressing. *Communications of the ACM*, 17(7):403–412, 1974. DOI: 10.1145/361011.361070

[90] G. Faden. Reconciling CMW requirements with those of X11 applications. In *Proceedings of the 14th Annual National Computer Security Conference*, October 1991.

[91] R. Feiertag. A technique for proving specifications are multilevel secure. Technical Report CSL-109, Stanford Research Institute, 1980.

[92] R. J. Feiertag and P. G. Neumann. The foundations of a provably secure operating system (PSOS). In *Proceedings of the National Computer Conference*, pp. 329–334, 1979.

[93] J. S. Fenton. Memoryless subsystems. *Computer Journal*, 17(2):143–147, May 1974. DOI: 10.1093/comjnl/17.2.143

[94] D. F. Ferraiolo and D. R. Kuhn. Role-based access control. In *15th National Computer Security Conference*, October 1992.

[95] N. Feske and C. Helmuth. A nitpicker's guide to a minimal-complexity secure GUI. In *Proceedings of the 21st Annual Computer Security Applications Conference*, pp. 85–94, 2005. DOI: 10.1109/CSAC.2005.7

[96] T. Fine and S. E. Minear. Assuring Distributed Trusted Mach. In *Proceedings of the IEEE Symposium on Security and Privacy*, pp. 206–218, 1993. DOI: 10.1109/RISP.1993.287631

[97] Secure minicomputing operating system (KSOS) executive summary: Phase I: Design of the department of defense kernelized secure operating system. Technical Report WDL-781, Ford Aerospace and Communications Corp. Available at http://csrc.nist.gov/publications/history/ford78.pdf, March 1978.

[98] B. Ford, M. Hibler, J. Lepreau, P. Tullmann, G. Back, and S. Clawson. Microkernels meet recursive virtual machines. In *Proceedings of the 2nd Symposium on Operating Systems Design and Implementation*, pp. 137–151, 1996. DOI: 10.1145/238721.238769

[99] L. J. Fraim. SCOMP: A solution to the multilevel security problem. *IEEE Computer*, 16(7):26–34, 1983. DOI: 10.1109/TC.1983.1676120

[100] L. J. Fraim. Secure office management system: The first commodity application on a trusted system. In *Proceedings of the 1987 Fall Joint Computer Conference on Exploring TEchnology: Today and Tomorrow*, pp. 421–426, 1987.

[101] T. Fraser. LOMAC: Low water-mark integrity protection for COTS environments. In *Proceedings of the 2000 IEEE Symposium on Security and Privacy*, May 2000. DOI: 10.1109/SECPRI.2000.848460

[102] T. Fraser, L. Badger, and M. Feldman. Hardening COTS software with generic software wrappers. In *Proceedings of the 1999 IEEE Symposium on Security and Privacy*, May 1999. DOI: 10.1109/SECPRI.1999.766713

[103] The FreeBSD Project. http://www.freebsd.org/, 2008.

[104] K. Fu, M. F. Kaashoek, and D. Mazières. Fast and secure distributed read-only file system. *ACM Transactions on Computer Systems*, 20(1):1–24, 2002. DOI: 10.1145/505452.505453

[105] T. Garfinkel, B. Pfaff, J. Chow, M. Rosenblum, and D. Boneh. Terra: A virtual machine-based platform for trusted computing. In *Proceedings of the 19th ACM Symposium on Operating System Principles (SOSP 2003)*, Bolton Landing, NY, October 2003. DOI: 10.1145/945445.945464

[106] T. Garfinkel and M. Rosenblum. A virtual machine introspection-based architecture for intrusion detection. In *Proceedings of the 2003 ISOC Symposium on Networked and Distributed System Security Symposium (NDSS'03)*, San Diego, CA, February 2003.

[107] T. Garfinkel and M. Rosenblum. When virtual is harder than real: Security challenges in virtual machine based computing environments. In *Proceedings of the 10th Workshop on Hot Topics in Operating Systems*, May 2005.

[108] M. Gasser. *Building a Secure Computer System*. Van Nostrand Reinhold. http://cs.unomaha.edu/~stanw/gasserbook.pdf, 1988.

[109] Gemini Trusted Network Processor—Class A1 Evaluation. Available at http://www.aesec.com/eval/CSC-EPL-94-008.html.

[110] C. Girling. Covert channels in LANs. *IEEE Transactions on Software Engineering*, 13(2):292–296, February 1987. DOI: 10.1109/TSE.1987.233153

[111] V. D. Gligor, C. S. Chandersekaran, R. S. Chapman, L. J. Dotterer, M. S. Hecht, W-D. Jiang, A. Johri, G. L. Luckenbaugh, and N. Vasudevan. Design and implementation of Secure Xenix. *IEEE Transactions on Software Engineering*, 13(2):208–221, 1987. DOI: 10.1109/TSE.1987.232893

[112] GPL General Public License. http://www.gnu.org/licenses/gpl.html, 2007.

[113] J. Goguen and J. Meseguer. Security policies and security models. In *Proceedings of the 1982 IEEE Symposium on Security and Privacy*, 1982.

[114] B. D. Gold, R. R. Linde, and P. F. Cudney. KVM/370 in retrospect. In *Proceedings of the 1984 IEEE Symposium on Security and Privacy*, pp. 13–23, May 1984. DOI: 10.1109/SP.1984.10002

[115] I. Goldberg, D. Wagner, R. Thomas, and E. A. Brewer. A secure environment for untrusted helper applications. In *Proceedings of the 6th Usenix Security Symposium*, San Jose, CA, 1996.

[116] D. B. Golub, R. W. Dean, A. Forin, and R. F. Rashid. UNIX as an application program. In *Proceedings of the 1990 USENIX Summer Conference*, pp. 87–95, 1990.

[117] L. Gong, G. Ellison, and M. Dageforde. *Inside Java 2 Platform Security*. Addison-Wesley, 2003.

[118] D. I. Good. Mechanical proofs about computer programs. In *Mathematical Logic and Programming Languages*. Prentice-Hall, 1985.

[119] B. Goodheart and J. Cox. *The Magic Garden Explained: The Internals of Unix System V*. Prentice-Hall, 1994.

[120] S. Govindavajhala and A. W. Appel. Windows Access Control Demystified. http://www.cs.princeton.edu/~sudhakar/papers/winval.pdf, January 2006.

[121] Homepage of PaX. http://pax.grsecurity.net/, 2008.

[122] S. Gupta and V. D. Gligor. Towards a theory of penetration-resistant systems and its applica-tion. In *Proceedings of the 4th IEEE Computer Security Foundations Workshop*, pp. 62–78, June 1991. DOI: 10.1109/CSFW.1991.151571

[123] S. Gupta and V. D. Gligor. Experience with a penetration analysis method and tool. In *Proceedings of the 15th National Computer Security Conference*, pp. 165–183, October 1992.

[124] J. T. Haigh, R. A. Kemmerer, J. McHugh, and W. D. Young. An experience using two covert channel analysis techniques on a real system design. *IEEE Transactions on Software Engineer-ing*, 13(2):157–168, 1987. DOI: 10.1109/TSE.1987.226479

[125] J. T. Haigh and W. D. Young. Extending the noninterference version of MLS for SAT. *IEEE Transactions on Software Engineering*, 13(2):141–150, 1987. DOI: 10.1109/TSE.1987.226478

[126] T. Haigh. Multicians.org and the history of operating systems. http://www.cbi.umn.edu/iterations/haigh.html, Sept 2002.

[127] S. E. Hallyn and P. Kearns. Domain and type enforcement for Linux. In *Proceedings of the 4th Annual Linux Showcase and Conference.* At http://www.sagecertification.org/publications/library/proceedings/als0 0/2000papers/papers/full_papers/hallyn/hallyn_html/index.html, October 2000.

[128] N. Hardy. The KeyKOS architecture. *Operating Systems Review*, 19(4):8–25, October 1985. DOI: 10.1145/858336.858337

[129] N. Hardy. The confused deputy. *Operating Systems Review*, 22(4):36–38, October 1988. DOI: 10.1145/54289.871709

[130] M. Harrison, W. Ruzzo, and J. D. Ullman. Protection in operating systems. *Communications of the ACM*, August 1976. DOI: 10.1145/360303.360333

[131] W. S. Harrison, N. Hanebutte, P. Oman, and J. Alves-Foss. The MILS architecture for a secure global information grid. *Crosstalk: The Journal of Defense Software Engineering*, 10(10):20–24, October 2005.

[132] G. Heiser, K. Elphinstone, I Kuz, G. Klein, and S. M. Petters. Towards trustworthy computing systems: Taking microkernels to the next level. *ACM Operating Systems Review*, 41(4):3–11, July 2007. DOI: 10.1145/1278901.1278904

[133] G. Heiser, K. Elphinstone, J. Vochteloo, S. Russell, and J. Liedtke. The Mungi single-address-space operating system. *Software Practice and Experience*, 18(9), July 1998. DOI: 10.1002/(SICI)1097-024X(19980725)28:9<901::AID-SPE181>3.0.CO;2-7

[134] C. Heitmeyer, M. Archer, E. Leonard, and J. McLean. Formal specification and verification of data separation in a separation kernel for an embedded system. In *Proceedings of the 13th ACM Conference on Computer and Communications Security*, pp. 346–355, 2006. DOI: 10.1145/1180405.1180448

[135] A. J. Herbert. A new protection architecture for the Cambridge Capability Computer. *ACM SIGOPS Operating Systems Review*, 12(1), 1978. DOI: 10.1145/775323.775326

[136] HP NetTop: A Technical Overview. Available at: `http://h71028.www7.hp.com/enterprise/downloads/HP_NetTop_Whitepaper2.pdf`, 2004.

[137] G. Hoglund and J. Butler. *Rootkits: Subverting the Windows Kernel*. Addison-Wesley, 2005.

[138] M. E. Houdek, F. G. Soltis, and R. L. Hoffman. IBM System/38 support for capability-based addressing. In *Proceedings of 8th Symposium on Computer Architecture*, pp. 341–348, May 1981.

[139] M. Howard and S. Lipner. *The Security Development Lifecycle*. Microsoft Press, 2006.

[140] W-M. Hu. Reducing timing channels with fuzzy time. In *Proceedings of the 1991 IEEE Symposium on Security and Privacy*, May 1991. DOI: 10.1109/RISP.1991.130768

[141] J. Humphreys and T. Grieser. Mainstreaming server virtualization: The Intel approach. Technical report, Intel Corporation. `http://www.intel.com/business/technologies/idc_virtualization_wp.pdf`, June 2006.

[142] System z PR/SM. `http://publib.boulder.ibm.com/infocenter/eserver/v1r2/index.jsp?topic=/eicaz/eicazzlpar.htm`, 2007.

[143] IBM z/VM Operating System. International Business Machines Corp. at `http://www.vm.ibm.com/`, 2008.

[144] J. W. Gray III. Probabilistic interference. In *Proceedings of the 1990 IEEE Symposium on Security and Privacy*, 1990. DOI: 10.1109/RISP.1990.63848

[145] C. Irvine. The reference monitor concept as a unifying principle in computer security education. In *Proceedings of the 1st World Conference on Information Systems Security Education*, June 1999.

[146] C. E. Irvine. A multilevel file system for high assurance. In *Proceedings of the 1995 IEEE Symposium on Security and Privacy*, 1995. DOI: 10.1109/SECPRI.1995.398924

[147] Information Technology Security Evaluation Criteria. Commission of the European Communities, June 1991.

[148] T. Jaeger, K. Butler, D. H. King, S. Hallyn, J. Latten, and X. Zhang. Leveraging IPsec for mandatory access control across systems. In *Proceedings of the Second International Conference on Security and Privacy in Communication Networks*, August 2006. DOI: 10.1109/SECCOMW.2006.359530

[149] T. Jaeger, A. Edwards, and X. Zhang. Consistency analysis of authorization hook placement in the Linux security modules framework. *ACM Transactions on Information and System Security (TISSEC)*, 7(2):175–205, May 2004. DOI: 10.1145/996943.996944

[150] T. Jaeger, R. Sailer, and X. Zhang. Analyzing integrity protection in the SELinux example policy. In *Proceedings of the 12th USENIX Security Symposium*, pp. 59–74, August 2003.

[151] P. A. Janson. *Using Type Extension to Organize Virtual Machine Mechanisms*. PhD thesis, Massachusetts Institute of Technology, September 1976.

[152] J. Johansson. Security watch: The long-term impact of user account control. http://technet.microsoft.com/en-us/magazine/cc137811.aspx, 2008.

[153] M. St. Johns. Draft revised IP security option. IETF RFC 1038.

[154] R. Johnson and D. Wagner. Finding user/kernel pointer bugs with type inference. In *Proceedings of the 13th conference on USENIX Security Symposium*, pp. 9–9, 2004.

[155] A. Jones, R. Lipton, and L. Snyder. A linear time algorithm for deciding security. In *Proceedings of the 17th Annual Symposium on Foundations of Computer Science*, 1976.

[156] P-H. Kamp and R. N. M. Watson. Jails: Confining the omnipotent root. Available at http://docs.freebsd.org/44doc/papers/jail/jail.html. Originally, presented in the 2nd International SANE Conference, 2000.

[157] P. A. Karger. *Improving Security and Performance for Capability Systems*. PhD thesis, University of Cambridge, October 1988.

[158] P. A. Karger and A. J. Herbert. An augmented capability architecture to support lattice security and traceability of access. In *Proceedings of the 1984 IEEE Symposium on Security and Privacy*, pp. 2–12, May 1984. DOI: 10.1109/SP.1984.10001

[159] P. A. Karger and R. R. Schell. MULTICS security evaluation: Vulnerability analysis. Technical Report ESD-TR-74-193, Deputy for Command and Management Systems, Electronics Systems Division (ASFC), L. G. Hanscom Field, Bedford, MA, June 1974. Reprinted in the Proceedings of the 2004 Annual Computer Security Applications Conference.

[160] P. A. Karger and J. C. Wray. Storage channels in disk arm optimization. In *Proceedings of the 1991 IEEE Symposium on Security and Privacy*, May 1991. DOI: 10.1109/RISP.1991.130771

[161] P. A. Karger, M. E. Zurko, D. W. Bonin, A. H. Mason, and C. E. Kahn. A retrospective on the VAX VMM security kernel. *IEEE Transactions on Software Engineering*, 17(11):1147–1165, 1991. DOI: 10.1109/32.106971

[162] R. Kemmerer. A practical approach to identifying storage and timing channels. *Proceedings of the 1982 IEEE Symposium on Security and Privacy*, 1982.

[163] R. A. Kemmerer. Shared Resource Matrix methodology: An approach to identifying storage and timing channels. *ACM Transactions on Computer Systems*, 1(3):256–277, 1983. DOI: 10.1145/357369.357374

[164] R. A. Kemmerer and P. A. Porras. Covert Flow Trees: A visual approach to analyzing covert storage channels. *IEEE Transactions on Software Engineering*, 17(11):1166–1185, 1991. DOI: 10.1109/32.106972

[165] S. Kent. Security options for the internet protocol. IETF RFC 1108.

[166] S. Kent and R. Atkinson. IP authentication header. IETF RFC 2402.

[167] S. Kent and R. Atkinson. IP encapsulating security payload. IETF RFC 2406.

[168] S. Kent and R. Atkinson. Security architecture for the internet protocol. IETF RFC 2401. DOI: 10.1016/S0167-4048(97)81995-5

[169] G. H. Kim and E. H. Spafford. The design and implementation of Tripwire: A file system integrity checker. In *Proceedings of the 2nd ACM Conference on Computer and Communications Security (CCS)*, pp. 18–29, 1994. DOI: 10.1145/191177.191183

[170] S. T. King, P. M. Chen, Y-M. Wang, C. Verbowski, H. J. Wang, and J. R. Lorch. SubVirt: Implementing malware with virtual machines. In *Proceedings of the 2006 IEEE Symposium on Security and Privacy*, May 2006.

[171] C. Ko, T. Fraser, L. Badger, and D. Kilpatrick. Detecting and countering system intrusions using software wrappers. In *Proceedings of the 9th USENIX Security Symposium*, 2000. DOI: 10.1109/SP.1984.10000

[172] S. Kramer. The MITRE flow table generator. Technical Report M83-31, The MITRE Corporation, January 1983.

[173] S. Kramer. Linus IV-An experiment in computer security. In *Proceedings of the 1984 IEEE Symposium on Security and Privacy*, 1984.

[174] M. N. Krohn, A. Yip, M. Brodsky, N. Cliffer, M. F. Kaashoek, E. Kohler, and R. Morris. Information flow control for standard OS abstractions. In *Proceedings of the 21st ACM Symposium on Operating Systems Principles*, pp. 321–334, October 2007. DOI: 10.1145/1294261.1294293

[175] M. Lageman. Solaris Containers—What They Are and How to Use Them. http://www.sun.com/blueprints/0505/819-2679.pdf, May 2005.

[176] B. W. Lampson. Protection. In *5th Princeton Conference on Information Sciences and Systems*, 1971.

[177] B. W. Lampson. A note on the confinement problem. *Communications of the ACM*, 16(10):613–615, 1973. DOI: 10.1145/362375.362389

[178] B. W. Lampson and H. E. Sturgis. Reflections on an operating system design. *Communications of the ACM*, 19(5):251–265, May 1976. DOI: 10.1145/360051.360074

[179] C. E. Landwehr. Formal models for computer security. *ACM Computing Surverys*, 13(3):247–278, 1981. DOI: 10.1145/356850.356852

[180] C. E. Landwehr. The best available technologies for computer security. *IEEE Computer*, 16(7):86–100, 1983. DOI: 10.1109/MC.1983.1654445

[181] H. M. Levy. *Capability-Based Computer Systems*. Digital Press. Available at http://www.cs.washington.edu/homes/levy/capabook/, 1984.

[182] N. Li, Z. Mao, and H. Chen. Usable mandatory integrity protection for operating systems. In *Proceedings of the 2007 IEEE Symposium on Security and Privacy*, May 2007. DOI: 10.1109/SP.2007.37

[183] LIDS Secure Linux System. http://www.lids.org/, 2008.

[184] J. Liedtke. Improving IPC by kernel design. In *Proceedings of the Fourteenth ACM Symposium on Operating Systems Principles*, pp. 175–188, 1993. DOI: 10.1145/168619.168633

[185] J. Liedtke, K. Elphinstone, S. Schonberg, H. Haertig, G. Heiser, N. Islam, and T. Jaeger. Achieved IPC performance. In *Proceedings of the 1997 Workshop on Hot Topics in Operating Systems*, pp. 28–31, 1997. DOI: 10.1109/HOTOS.1997.595177

[186] T. A. Linden. Operating system structures to support security and reliable software. *ACM Computing Surveys*, 8(4):409–445, December 1976. DOI: 10.1145/356678.356682

[187] The Linux kernel archives. http://www.kernel.org/, 2008.

[188] Kernel Summit 2006 - Security. http://lwn.net/Articles/191737/, July 2006.

[189] S. B. Lipner. A comment on the confinement problem. In *Proceedings of the Fifth ACM Symposium on Operating Systems*, 1975. DOI: 10.1145/1067629.806537

[190] S. B. Lipner. Non-discretionery controls for commercial applications. In *Proceedings of the 1982 IEEE Symposium on Security and Privacy*, 1982. DOI: 10.1109/SP.1982.10022

[191] P. A. Loscocco, S. D. Smalley, P. A. Muckelbauer, R. C. Taylor, S. J. Turner, and J. F. Farrell. The Inevitability of Failure: The flawed assumption of security in modern computing environments. In *Proceedings of the 21st National Information Systems Security Conference*, pp. 303–314, October 1998.

[192] R. Love. *Linux Kernel Development*. Sams, 2003.

[193] W. Martin, P. White, F. Taylor, and A. Goldberg. Formal construction of the mathematically analyzed separation kernel. In *Proceedings of the 15th International Conference on Automated Software Engineering*, pp. 133–141, 2001. DOI: 10.1109/ASE.2000.873658

[194] E. Mate-Bacic. The Canadian Trusted Computer Product Evaluation Criteria. In *Proceedings of the 1990 Annual Computer Security Applications Conference*, pp. 188–196, 1990. DOI: 10.1109/CSAC.1990.143768

[195] F. Mayer, K. Macmillan, and D. Caplan. *SELinux by Example: Using Security-Enhanced Linux*. Addison-Wesley, 2006.

[196] F. L. Mayer. An interpretation of a refined Bell-La Padula model for the TMach kernel. In *Proceedings of the 1988 Aerospace Computer Security Applications Conference*, pp. 368–378, December 1988. DOI: 10.1109/ACSAC.1988.113335

[197] D. Mazières, M. Kaminsky, M. F. Kaashoek, and E. Witchel. Separating key management from file system security. In *Proceedings of the 17th ACM Symposium on Operating System Principles*, pp. 124–139, 1999. DOI: 10.1145/319344.319160

[198] E. J. McCauley and P. J. Drongowski. KSOS: The design of a secure operating system. In *Proceedings of the 1979 National Computer Conference*, vol. 48, pp. 345–353, 1979.

[199] D. McIlroy and J. Reeds. Multilevel windows on a single-level terminal. In *Proceedings of the (First) USENIX Security Workshop*, August 1988.

[200] M. D. McIlroy and J. A. Reeds. Multilevel security in the UNIX tradition. *Software—Practice and Experience*, 22:673–694, 1992. DOI: 10.1002/spe.4380220805

[201] M. K. McKusick, K. Bostic, M. J. Karels, and J. S. Quarterman. *The Design and Implementation of the 4.4BSD Operating System*. Addison-Wesley, 1996.

[202] J. McLean. Proving noninterference and functional correctness using traces. *Journal of Computer Security*, 1(1), 1992.

[203] J. McLean. Security models. In J. Marciniak, editor, *Encyclopedia of Software Engineering*. John Wiley & Sons, 1994.

[204] Medusa DS9 Security System. http://medusa.terminus.sk/, 2008.

[205] R. Meushaw and D. Simard. NetTop: Commercial technology in high assurance applications. *Tech Trend Notes*, 9(4):1–8, 2000.

[206] *Microsoft Computer Dictionary*. Microsoft Press, fifth edition, 2002.

[207] Microsoft Developer Network. Microsoft Library at http://msdn.microsoft.com/en-us/library/default.aspx, 2008.

[208] Microsoft Virtual PC. Microsoft document at http://www.microsoft.com/windows/products/winfamily/virtualpc/default.mspx, 2008.

[209] Inside the Secure Windows Initiative. http://www.microsoft.com/technet/archive/security/bestprac/secwinin.mspx?mfr=true.

[210] PREfast for Drivers. http://www.microsoft.com/whdc/devtools/tools/prefast.mspx.

[211] J. K. Millen. Covert channel capacity. In *Proceedings of the 1987 IEEE Symposium on Security and Privacy*, 1987.

[212] J. K. Millen. 20 years of covert channel modeling and analysis. In *Proceedings of the 1999 IEEE Symposium on Security and Privacy*, pp. 113–114, 1999.

[213] S. E. Minear. Providing policy control over object operations in a Mach-Based system. In *Proceedings of the 5th USENIX Security Symposium*, pp. 141–156, 1995.

[214] P. Moore. NetLabel – Explicit labeled networking for Linux. http://netlabel.sourceforge.net/, October 2007.

[215] I. S. Moskowitz and A. R. Miller. The channel capacity of a certain noisy timing channel. *IEEE Transactions on Information Theory*, 38(4):1339–1344, 1992. DOI: 10.1109/18.144712

[216] S. Mullender. *Principles of Distributed Operating System Design*. PhD thesis, Vrije University, October 1985.

[217] Multics history. http://www.multicians.org/history.html, Apr 2008.

[218] A. C. Myers and B. Liskov. A decentralized model for information flow control. In *Proceedings of the 16th ACM Symposium on Operating System Principles*, October 1997. DOI: 10.1145/268998.266669

[219] A. C. Myers, N. Nystrom, L. Zheng, and S. Zdancewic. Jif: Java + information flow. Software release. Located at http://www.cs.cornell.edu/jif, July 2001.

[220] R. Naraine. Hacker, Microsoft duke it out over Vista design flaw.
http://blogs.zdnet.com/security/?p=29, February 2007.

[221] R. Naraine. Russinovich: Malware will thrive, even with Vista's UAC.
http://blogs.zdnet.com/security/?p=175, April 2007.

[222] Rainbow series. http://en.wikipedia.org/wiki/Rainbow_Series, 2008.

[223] R. M. Needham and R. Walker. The Cambridge CAP computer and its protection system.
In *Proceedings of the 6th ACM Symposium on Operating System Principles*, pp. 1–10, 1977.
DOI: 10.1145/1067625.806541

[224] The NetBSD Project. http://www.netbsd.org/, 2008.

[225] B. C. Neuman and T. Ts'o. Kerberos: An authentication service for computer networks. *IEEE Communications*, 32(9):33–38, 1994. DOI: 10.1109/35.312841

[226] P. G. Neumann, R. S. Boyer, R. J. Feiertag, K. N. Levitt, and L. Robinson. A provably secure operating system: The system, its applications, and proofs. Technical Report CSL-116, Stanford Research Institute, May 1980.

[227] A. Noodergraaf and K. Watson. Solaris Operating Environment Security.
http://www.sun.com/blueprints/0100/security.pdf, January 2000.

[228] AppArmor Linux application security.
http://www.novell.com/linux/security/apparmor/, 2008.

[229] Security-Enhanced Linux. http://www.nsa.gov/selinux.

[230] Controlled Access Protection Profile, Version 1.d.
http://www.commoncriteriaportal.org/files/ppfiles/capp.pdf, October 1999.

[231] Labeled Security Protection Profile, Version 1.b.
http://www.commoncriteriaportal.org/files/ppfiles/lspp.pdf, October 1999.

[232] A. One. Smashing the stack for fun and profit. *Phrack*, 7(49), Available at
http://www.phrack.org/issues.html?id=14&issue=49, 1997.

[233] The Partitioning Kernel Protection Profile. The Open Group, June 2003. Draft Under Review.

[234] OpenAFS. http://www.openafs.org/, 2008.

[235] OpenBSD. http://www.openbsd.org/, 2008.

[236] OpenWall Project - Information security software for open environments.
http://www.openwall.com/, 2008.

[237] E. Organick. *The Multics System: An Examination of its Structure*. MIT Press, http://www.multicians.org/flass-organick.html, 1972.

[238] E. Organick. *A Programmer's View of the Intel 432 System*. McGraw-Hill, 1983.

[239] A. Ott. RSBAC and LSM. http://www.rsbac.org/documentation/why_rsbac_does_not_use_lsm, 2006.

[240] A. Ott. RSBAC: Extending Linux security beyond the limits. http://www.rsbac.org/, 2008.

[241] D. L. Parnas. On the criteria to be used in decomposing systems into modules. *Communications of the ACM*, 15(12):1053–1058, 1972. DOI: 10.1145/361598.361623

[242] D. Paul. A summary of the Unisys experience with GEMSOS. In *Proceedings of the 1989 Annual Computer Security Applications Conference*, pp. 112–113, 1989. DOI: 10.1109/CSAC.1989.81039

[243] Vista backlash: Microsoft quietly lets Vista users revert to XP. http://blogs.pcworld.com/staffblog/archives/005512.html, September 2007.

[244] N. L. Petroni, Jr., T. Fraser, J. Molina, and W. A. Arbaugh. Copilot - a coprocessor-based kernel runtime integrity monitor. In *Proceedings of the 13th USENIX Security Symposium*, pp. 179–194, 2004.

[245] N. L. Petroni, Jr., T. Fraser, A. Walters, and W. A. Arbaugh. An architecture for specification-based detection of semantic integrity violations in kernel dynamic data. In *Proceedings of the 15th USENIX Security Symposium*, 2006.

[246] G. Popek and D. Farber. A model for the verification of data security in operating systems. *Communications of the ACM*, 21(9):237–249, September 1978. DOI: 10.1145/359588.359597

[247] G. J. Popek and R. P. Goldberg. Formal requirements for virtualizable third generation architectures. *Communications of the ACM*, 17(7):412–421, July 1974. DOI: 10.1145/361011.361073

[248] G. J. Popek, A. Kampe, C. S. Kline, A. Stoughton, M. Urban, and E. J. Walton. UCLA Secure Unix. In *Proceedings of the AFIPS National Computer Conference*, vol. 48, pp. 355–364, 1979.

[249] N. E. Proctor and P. G. Neumann. Architectural implications of covert channels. In *Proceedings of the Fifteenth National Computer Security Conference*, pp. 28–43. http://www.csl.sri.com/users/neumann/ncs92.html, October 1992.

[250] N. Provos. Improving host security with system call policies. In *Proceedings of the 2003 USENIX Security Symposium*, August 2003.

[251] N. Provos, M. Friedl, and P. Honeyman. Preventing privilege escalation. In *Proceedings of the USENIX Security Symposium*, August 2003.

[252] D. Redell. *Naming and Protection in Extendible Operating Systems*. PhD thesis, University of California, Berkeley, 1974. Reprinted as Project MAC TR-140, Massachusetts Institute of Technology.

[253] D. P. Reed. *Processor Multiplexing In A Layed Operating System*. PhD thesis, Massachusetts Institute of Technology, 1976.

[254] D. P. Reed. Processor multiplexing in a layered operating system. Technical Report MIT/LCS/TR-164, Massachusetts Institute of Technology, July 1976.

[255] P. Reiher, T. Page, S. Crocker, J. Cook, and G. Popek. Truffles—a secure service for widespread file sharing. In *Proceedings of the The PSRG Workshop on Network and Distributed System Security*, February 1993.

[256] J. Reynolds and R. Chandramouli. Role-Based Access Control Protection Profile, Version 1.0. http://www.commoncriteriaportal.org/files/ppfiles/RBAC_987.pdf, July 1998.

[257] R. L. Rivest, A. Shamir, and L. Adleman. A method for obtaining digital signatures and public-key cryptosystems. *Communications of the ACM*, 21(2):120–126, 1978. DOI: 10.1145/359340.359342

[258] J. S. Robin and C. E. Irvine. Analysis of the Intel Pentium's ability to support a secure virtual machine monitor. In *Proceedings of the 9th conference on USENIX Security Symposium*, 2000.

[259] J. Rushby. Design and verification of secure systems. In *Proceedings of the Eighth ACM Symposium on Operating System Principles*, pp. 12–21, December 1981. DOI: 10.1145/800216.806586

[260] J. Rushby. Proof of separability: A verification technique for a class of security kernels. In *Proceedings of the International Symposium on Programming*, pp. 352–367, 1982.

[261] J. Rushby. Noninterference, transitivity and channel-control security policies. Technical Report CSL-92-02, Stanford Research Institute, December 1992.

[262] Best practices for UNIX chroot() operations. http://www.unixwiz.net/techtips/chroot-practices.html.

[263] R. Sailer, T. Jaeger, E. Valdez, R. Cáceres, R. Perez, S. Berger, J. Griffin, and L. van Doorn. Building a MAC-based security architecture for the Xen opensource hypervisor. In *Proceedings of the 21st Annual Computer Security Applications Conference (ACSAC 2005)*, Miami, FL, December 2005. DOI: 10.1109/CSAC.2005.13

[264] J. H. Saltzer. Protection and the control of information sharing in Multics. *Communications of the ACM*, 17(7):388–402, July 1974. DOI: 10.1145/361011.361067

[265] J. H. Saltzer and M. D. Schroeder. The protection of information in computer systems. *Proceedings of the IEEE*, 63(9), September 1975. DOI: 10.1109/PROC.1975.9939

[266] P. H. Salus. *A Quarter Century of UNIX*. Addison-Wesley, 1994.

[267] R. Sandberg, D. Goldberg, S. Kleiman, D. Walsh, and B. Lyon. Design and implementation of the Sun Network Filesystem. In *Proceedings of the 1985 Summer USENIX Conference*, pp. 119–130, 1985.

[268] R. Sandhu, D. Ferraiolo, and R. Kuhn. The NIST Model for Role-Based Access Control: Towards a unified standard. In *Proceedings of the 5th ACM Role-Based Access Control Workshop*, July 2000. DOI: 10.1145/344287.344301

[269] R. S. Sandhu. The Schematic Protection Model: Its definition and analysis for acyclic attenuating schemes. *Journal of the ACM*, 35(2):404–432, 1988. DOI: 10.1145/42282.42286

[270] R. S. Sandhu. The Typed Access Matrix model. In *Proceedings of the 1992 IEEE Symposium on Security and Privacy*, 1992. DOI: 10.1109/RISP.1992.213266

[271] R. S. Sandhu. Lattice-based access control models. *IEEE Computer*, 26(11):9–19, 1993. DOI: 10.1109/2.241422

[272] R. S. Sandhu, E. J. Coyne, H. L. Feinstein, and C. E. Youman. Role-Based Access Control Models. *IEEE Computer*, 29(2):38–47, 1996.

[273] O. S. Saydjari, J. K. Beckman, and J. R. Leaman. LOCKing computers securely. In *Proceedings of the 10th National Computer Security Conference*, pp. 129–141, http://stinet.dtic.mil/cgi-bin/GetTRDoc?AD=ADA219100&Location=U2&doc=GetTRDoc.pdf, 1987.

[274] O. S. Saydjari, J. K. Beckman, and J. R. Leaman. LOCK Trek: Navigating uncharted space. In *Proceedings of the 1989 IEEE Symposium on Security and Privacy*, 1989. DOI: 10.1109/SECPRI.1989.36291

[275] M. Schaefer, B. Gold, R. Linde, and J. Scheid. Program confinement in KVM/370. In *Proceedings of the 1977 ACM Annual Conference*, pp. 404–410, October 1977. DOI: 10.1145/800179.1124633

[276] M. A. Schaffer and G. Walsh. LOCK/ix: On implementing Unix on the LOCK TCB. In *Proceedings of the 11th National Computer Security Conference*, 1988.

[277] R. Schell, T. Tao, and M. Heckman. Designing the GEMSOS security kernel for security and performance. In *Proceedings of the National Computer Security Conference*, 1985.

[278] G. Schellhorn, W. Reif, A. Schairer, P. A. Karger, V. Austel, and D. Toll. Verification of a formal security model for multiapplicative smart cards. In *Proceedings of the European Symposium on Research in Computer Security*, pp. 17–36, 2000.

[279] M. D. Schroeder. Engineering a security kernel for Multics. In *Proceedings of the Fifth ACM Symposium on Operating Systems Principles*, pp. 25–32, 1975. DOI: 10.1145/800213.806518

[280] M. D. Schroeder, D. D. Clark, J. H. Saltzer, and D. Wells. Final report of the MULTICS kernel design project. Technical Report MIT-LCS-TR-196, MIT, March 1978. DOI: 10.1145/800214.806546

[281] M. D. Schroeder and J. H. Saltzer. A hardware architecture for implementing protection rings. *Communications of the ACM*, 15(3):157–170, 1972. DOI: 10.1145/361268.361275

[282] Software requirements specification for Distributed Trusted Mach. Technical Report DT-Mach CDRL A005, Secure Computing Corporation, June 1992.

[283] DTOS Lessons Learned Report. Technical Report DTOS CDRL A008, Secure Computing Corporation. At
http://www.cs.utah.edu/flux/fluke/html/dtos/HTML/final-docs/lessons.pdf,
June 1997.

[284] G. Shah, A. Molina, and M. Blaze. Keyboards and covert channels. In *Proceedings of the 15th USENIX Security Symposium*, August 2006.

[285] U. Shankar, T. Jaeger, and R. Sailer. Toward automated information-flow integrity verification for security-critical applications. In *Proceedings of the 2006 ISOC Networked and Distributed Systems Security Symposium (NDSS'06)*, San Diego, CA, February 2006.

[286] J. S. Shapiro. *EROS: A Capability System*. PhD thesis, University of Pennsylvania, 1999.

[287] J. S. Shapiro. Verifying the EROS confinement mechanism. In *Proceedings of the 2000 IEEE Symposium on Security and Privacy*, May 2000. DOI: 10.1109/SECPRI.2000.848454

[288] J. S. Shapiro, J. M. Smith, and D. J. Farber. EROS: A fast capability system. In *Proceedings of the 17th ACM Symposium on Operating System Principles*, December 1999.

[289] J. S. Shapiro, J. Vanderburgh, E. Northup, and D. Chizmadia. Design of the EROS trusted window system. In *Proceedings of the 13th conference on USENIX Security Symposium*, 2004.

[290] W. R. Shockley, T. F. Tao, and M. F. Thompson. An overview of the GEMSOS class A1 technology and application experience. In *Proceedings of the 11th National Computer Security Conference*, pp. 238–245, October 1988.

[291] V. Simonet. The Flow Caml System: Documentation and User's Manual. Technical Report 0282, Institut National de Recherche en Informatique et en Automatique (INRIA), July 2003. ©INRIA.

[292] R. E. Smith. Constructing a high assurance mail guard. In *Proceedings of the 17th National Computer Security Conference*, 1994.

[293] R. E. Smith. Cost profile of a highly assured, secure operating system. *ACM Transactions on Information Systems Security*, 4(1):72–101, 2001. DOI: 10.1145/383775.383778

[294] R. W. Smith and G. S. Knight. Predictable design of network-based covert channel communication systems. In *Proceedings of the 2008 IEEE Symposium on Security and Privacy*, 2008. DOI: 10.1109/SP.2008.26

[295] R. Spencer, S. Smalley, P. Loscocco, M. Hibler, D. Andersen, and J. Lepreau. The Flask architecture: System support for diverse security policies. In *Proceedings of the 8th USENIX Security Symposium*, pp. 123–139, August 1999.

[296] B. Spengler. grsecurity. http://www.grsecurity.net/, 2008.

[297] B. Spengler. grsecurity LSM. http://www.grsecurity.net/lsm.php, 2008.

[298] G. R. Stoneburner and D. A. Snow. The Boeing MLS LAN: Heading towards an INFOSEC security solution. In *Proceedings of the 12th National Computer Security Conference*, pp. 254–266, 1989.

[299] J. Sugerman, G. Venkitachalam, and B-H. Lim. Virtualizing I/O devices on VMware Workstation's hosted virtual machine monitor. In *Proceedings of the 2002 USENIX Annual Technical Conference*, pp. 1–14, 2001.

[300] W. Sun, R. Sekar, G. Poothia, and T. Karandikar. Practical proactive integrity protection: A basis for malware defense. In *Proceedings of the 2008 IEEE Symposium on Security and Privacy*, May 2008. DOI: 10.1109/SP.2008.35

[301] Sun Microsystems. Understanding Security Attributes Assigned to Computers. Under Chapter 7 in http://docs.sun.com/app/docs/doc/816-1048/.

[302] D. Sutherland. A model of information. In *Proceedings of the Ninth National Computer Security Conference*, 1986.

[303] M. M. Swift, A. Hopkins, P. Brundrett, C. Van Dyke, P. Garg, S. Chan, M. Goertzel, and G. Jensenworth. Improving the granularity of access control for Windows 2000. *ACM Transactions on Information and Systems Security*, 5(4):398–437, 2002. DOI: 10.1145/581271.581273

[304] Trusted Computer System Evaluation Criteria (Orange Book). Technical Report DoD 5200.28-STD, U.S. Department of Defense, December 1985.

[305] How to exploit a format string vulnerability.
http://doc.bughunter.net/format-string/exploit-fs.html, 2008.

[306] M. F. Thompson, R. R. Schell, A. Tao, and T. Levin. Introduction to the Gemini Trusted Network Processor. In *Proceedings of the 13th National Computer Security Conference*, pp. 211–217, 1990.

[307] B. Tobotras. Linux kernel capabilities FAQ.
http://ftp.kernel.org/pub/linux/libs/security/linux-privs/kernel-2.4/capfaq-0.2.txt, April 1999.

[308] Policy management server. http://oss.tresys.com/projects/policy-server, Tresys Corp., 2008.

[309] Reference Policy. http://oss.tresys.com/projects/refpolicy, Tresys Corp., 2008.

[310] L. Tan, X. Zhang, X. Ma, W. Xiong, and Y. Zhou. AutoISES: Automatically inferring security specifications and detecting violations. In *Proceedings of the 17th USENIX Security Symposium*, 2008.

[311] J. T. Trostle. Modeling a fuzzy time system. In *Proceedings of the 1993 IEEE Symposium on Security and Privacy*, May 1993. DOI: 10.1109/RISP.1993.287641

[312] TrustedBSD - Home. http://www.trustedbsd.org/, 2008.

[313] The Distributed Trusted Operating System (DTOS) Home Page. At
http://www.cs.utah.edu/flux/fluke/html/dtos/HTML/dtos.html.

[314] A. Vahdat. *Operating System Services for Wide-Area Applications*. PhD thesis, University of California, Berkeley, December 1998.

[315] T. van Vleck. Timing channels. http://www.multicians.org/timing-chn.html, 1990.

[316] S. Vandebogart, P. Efstathopoulos, E. Kohler, M. N. Krohn, C. Frey, D. Ziegler, M. F. Kaashoek, R. Morris, and D. Mazières. Labels and event processes in the Asbestos operating system. *ACM Transactions on Computer Systems*, 25(4), 2007. DOI: 10.1145/1314299.1314302

[317] W. Venema. The Postfix home page. http://www.postfix.org.

[318] J. Viega and G. McGraw. *Building Secure Software: How to Avoid Security Problems the Right Way*. Addison-Wesley, 2001.

[319] VMware ESX bare-metal hypervisor for virtual machines. VMware Inc., document at
http://www.vmware.com/products/vi/esx/, 2008.

[320] VMware Server, virtual server consolidation, free virtualization. VMware Inc., document at http://www.vmware.com/products/server/, 2008.

[321] VMware VMsafe Security Technology. VMware Inc., document at http://www.vmware.com/overview/security/vmsafe/security_technology.html, 2008.

[322] V. A. Vyssotsky, F. J. Corbató, and R. M. Graham. Structure of the Multics supervisor. In *Proceedings of the 1965 AFIPS Fall Joint Computer Conference*, 1965.

[323] K. M. Walker, D. F. Sterne, M. L. Badger, M. J. Petkac, D. L. Sherman, and K. A. Oostendorp. Confining root programs with domain and type enforcement (DTE). In *Proceedings of the 6th USENIX Security Symposium*, 1996.

[324] D. Walsh. Using Reference Policy/Generating a reference policy module. http://danwalsh.livejournal.com/8707.html?thread=6451, 2007.

[325] E. Walsh. Application of the Flask architecture to the X Window System server. In *Proceedings of the 2007 SELinux Symposium*. Available at http://selinux-symposium.org/2007/agenda.php, March 2007.

[326] K. G. Walter. Primitive models for computer security. Technical report, Case Western Reserve University, January 1974.

[327] R. Watson, W. Morrison, C. Vance, and B. Feldman. The TrustedBSD MAC framework: Extensible kernel access control for FreeBSD 5.0. In *Proceedings of the USENIX Annual Technical Conference*, June 2003. DOI: 10.1109/DISCEX.2003.1194900

[328] R. N. M. Watson. TrustedBSD: Adding trusted operating system features to FreeBSD. In *Proceedings of the FREENIX Track: 2001 USENIX Annual Technical Conference*, pp. 15–28, 2001.

[329] C. Weissman. Security controls in the ADEPT-50 time-sharing system. In *Proceedings of the 1969 AFIPS Fall Joint Computer Conference*, pp. 119–133, 1969.

[330] C. Weissman. Blacker: Security for the DDN, examples of A1 security engineering trades. In *Proceedings of the 1992 IEEE Symposium on Security and Privacy*, pp. 286–292, May 1992. DOI: 10.1109/RISP.1992.213253

[331] T. Wheeler, S. Holtsberg, and S. Eckmann. Ina Go user's guide. Technical Report TM8613/003, Paramax Systems Corporation, 1992.

[332] A. Whitaker, M. Shaw, and S. D. Gribble. Scale and performance in the Denali isolation kernel. In *Proceedings of the 5th USENIX Symposium on Operating Systems Design and Implementation*, December 2002. DOI: 10.1145/1060289.1060308

[333] J. Whitmore, A. Bensoussan, P. Green, D. Hunt, A. Kobziar, and J. Stern. Design for MULTICS security enhancements. Technical Report ESD-TR-74-176, Honeywell Information Systems, Inc. Available at http://csrc.nist.gov/publications/history/whit74.pdf, December 1973.

[334] Code red (computer worm) - Wikipedia. http://en.wikipedia.org/wiki/Code_Red_worm.

[335] M. V. Wilkes. *Time-Sharing Computer Systems*. Elsevier Science Ltd., 3rd ed., 1975.

[336] M. V. Wilkes and R. M. Needham. *The Cambridge CAP Computer and Its Operating System*. Elsevier North Holland Inc., 1979.

[337] R. Wojtczuk. Defeating Solar Designer's non-executable stack patch. http://insecure.org/sploits/non-executable.stack.problems.html, 1998.

[338] M. Wolf. Covert channels in LAN protocols. In *Proceedings of the Workshop on Local Area Network Security*, 1991.

[339] R. M. Wong. A comparison of secure UNIX operating systems. In *Proceedings of the Sixth Annual Computer Security Applications Conference*, December 1990. DOI: 10.1109/CSAC.1990.143794

[340] J. P. L. Woodward. Security requirements for system high and compartmented mode workstation. Technical Report Document DDS-2600-5502-87, Defense Intelligence Agency, November 1987.

[341] J. C. Wray. An analysis of covert timing channels. In *Proceedings of the 1991 IEEE Symposium on Security and Privacy*, May 1991. DOI: 10.1109/RISP.1991.130767

[342] C. Wright, C. Cowan, S. Smalley, J. Morris, and G. Kroah-Hartman. Linux Security Modules: General security support for the Linux kernel. In *Proceedings of the 11th USENIX Security Symposium*, pp. 17–31, August 2002.

[343] W. Wulf, E. Cohen, W. Corwin, A. Jones, R. Levin, C. Pierson, and F. Pollack. Hydra: The kernel of a multiprocessor operating system. *Communications of the ACM*, 17(6), 1974. DOI: 10.1145/355616.364017

[344] W. A. Wulf, R. Levin, and S. P. Harbison. *HYDRA/C.mmp: An Experimental Computer System*. McGraw-Hill, 1981.

[345] The X Foundation: http://www.x.org.

[346] Xen Community. Available at http://xen.xensource.com/, 2008.

[347] V. Yodiaken. Response to Partitioning Kernel Protection Profile.
http://www.yodaiken.com/papers/securityx.pdf. FSMLabs Draft.

[348] M. Young, A. Tevanian, R. F. Rashid, D. B. Golub, J. L. Eppinger, J. Chew, W. J. Bolosky, D. L. Black, and R. V. Baron. The duality of memory and communication in the implementation of a multiprocessor operating system. In *Proceedings of the 11th Symposium on Operating Systems Principles*, pp. 63–76, 1987. DOI: 10.1145/37499.37507

[349] N. Zeldovich, S. Boyd-Wickizer, E. Kohler, and D. Mazières. Making information flow explicit in HiStar. In *Proceedings of the 7th Symposium on Operating System Design and Implementation*, pp. 263–278, 2006.

[350] N. Zeldovich, S. Boyd-Wickizer, and D. Mazières. Securing distributed systems with information flow control. In *Proceedings of the 5th Symposium on Networked Systems Design and Implementation*, April 2008.

[351] X. Zhang, A. Edwards, and T. Jaeger. Using CQUAL for static analysis of authorization hook placement. In *Proceedings of the 11th USENIX Security Symposium*, pp. 33–48, San Francico, CA, August 2002.

Biographies

Trent Jaeger is an Associate Professor in the Computer Science and Engineering Department at The Pennsylvania State University and the Co-Director of the Systems and Internet Infrastructure Security Lab. He joined Penn State after working for IBM Research for nine years in operating systems and system security research groups. Trent's research interests include operating systems security, access control, and source code and policy analysis tools. He has published over 70 refereed research papers on these subjects. Trent has made a variety of contributions to open source systems security, particularly to the Linux Security Modules framework, the SELinux module and policy development, integrity measurement in Linux, and the Xen security architecture. He is active in the security research community, having been a member of the program committees of all the major security conferences, and the program chair of the ACM CCS Government and Industry Track and ACM SACMAT, as well as chairing several workshops. He is an associate editor with ACM TOIT and has been a guest editor of ACM TISSEC. Trent has an M.S. and a Ph.D. from the University of Michigan, Ann Arbor, in Computer Science and Engineering in 1993 and 1997, respectively.

Glenn Faden is a Distinguished Engineer in the Solaris Security Technologies Group, and has worked at Sun for 19 years. He is currently the architect for Solaris Trusted Extensions, and was one of the architects for Trusted Solaris and Role-Based Access Control. He designed Sun's multilevel desktops based on Open Look, CDE, and GNOME; he holds a patent for the underlying X11 security policy. Glenn has made extensive contributions to the Solaris security foundation, including Access Control Lists, Auditing, Device Allocation, and OS Virtualization. He also developed the RBAC and Process Rights Management tools for the Solaris Management Console. He has authored several articles for Sun's Blueprints website, and the Solaris Developer Connection.

Glenn previously worked for Qubix, OmniCad, and Gould Computer Systems in Desktop Publishing and OS development. He has an MS in Computer Science from Florida Institute of Technology.

Christoph Schuba has studied mathematics and management information systems at the University of Heidelberg and the University of Mannheim in Germany. As a Fulbright scholar, he earned his M.S. and Ph.D. degrees in Computer Science from Purdue University in 1993 and 1997, performing most of his dissertation research in the Computer Science Laboratory at the Xerox Palo Alto Research Center (PARC). Christoph has taught undergraduate and graduate courses in computer and network security, cryptography, operating systems, and distributed systems at San José

State University, USA, at the Universtität Heidelberg, Germany, at the International University in Bruchsal, Germany, at Linköpings universitet in Linköping, Sweden, and most recently at Stanford University, USA.

Christoph has been working since 1997 at Sun Labs and most recently in the Solaris Software Security Organization at Sun Microsystems, Inc. He holds ten patents and is author and co-author of numerous scientific articles in computer and network security.

Index

Made in United States
North Haven, CT
29 July 2022

22002510R00130